普通高等教育规划教材

Practical English Listening Course for Pilots
飞行专业英语听力实用教程

杜丽娟　编著

北　京

冶金工业出版社

2021

内 容 提 要

本书分五讲，共计 18 课，主要内容涉及飞行历史，空气动力学，飞行原理，飞行仪表，初、高教机及民用运输机机型知识，飞机结构，飞行员面试及飞行训练，飞机主要系统及飞行中人的因素等。

本书可作为高等院校飞行技术、空中交通管制、飞行签派专业的教学用书，也可供航空企业、民航局及其下属管理局的相关工作人员阅读，并可供飞行爱好者参考。

本书配有音频材料，读者可扫描书中二维码收听。

图书在版编目(CIP)数据

飞行专业英语听力实用教程 = Practical English Listening Course for Pilots / 杜丽娟编著 . —北京：冶金工业出版社，2021. 1

普通高等教育规划教材

ISBN 978-7-5024-8682-2

Ⅰ.①飞… Ⅱ.①杜… Ⅲ.①航空—飞行术—英语—听说教学—高等学校—教材 Ⅳ.①V323

中国版本图书馆 CIP 数据核字(2021)第 016900 号

出 版 人　苏长永
地　　址　北京市东城区嵩祝院北巷 39 号　邮编　100009　电话　(010)64027926
网　　址　www. cnmip. com. cn　电子信箱　yjcbs@ cnmip. com. cn
责任编辑　王　颖　刘林烨　美术编辑　吕欣童　版式设计　禹　蕊
责任校对　郑　娟　责任印制　禹　蕊
ISBN 978-7-5024-8682-2
冶金工业出版社出版发行；各地新华书店经销；三河市双峰印刷装订有限公司印刷
2021 年 1 月第 1 版，2021 年 1 月第 1 次印刷
787mm×1092mm　1/16；11.25 印张；232 千字；172 页
49. 90 元
冶金工业出版社　投稿电话　(010)64027932　投稿信箱　tougao@cnmip. com. cn
冶金工业出版社营销中心　电话　(010)64044283　传真　(010)64027893
冶金工业出版社天猫旗舰店　yjgycbs. tmall. com
(本书如有印装质量问题，本社营销中心负责退换)

前　　言

当前，民航业对从业人员英语水平的要求逐年提高。飞行人员查阅飞行手册、飞行资料、陆空通话等都需要具备良好的英语水平。中国民航总局对国内飞行员的英语要求是与国际接轨的，各航空公司均把飞行员掌握专业英语的水平作为飞行员选拔的一项重要考核指标。鉴于此，编者精心编写了本书，旨在为提高我国飞行员英语水平贡献一份力量。

编者结合自身多年来从事飞行翻译及飞行员培训教学中积累的经验，查阅了大量的飞行专业英语资料，参考了国内外优秀的飞行专业英语教材，完成了本书的编写工作。本书分五讲，共计 18 课，并配有音频材料，读者可扫描书中二维码收听。主要内容涉及飞行历史，初、高教机及民用运输机机型知识，飞机主要系统及飞行中人的因素等。本书涉及中国民航发展史、中国大飞机事业的蓬勃发展等内容，在课程教学的过程中铺设思政元素，着力培养民航飞行事业的建设者和接班人，认真做到以树人为核心，坚定学生的理想信念，培养"准民航人"的家国情怀，教育引导他们将个人命运同祖国和民族的命运紧密连接，在奋斗与奉献中培育社会主义核心价值观，弘扬和践行当代民航精神，诠释对党和国家的深厚情感，敬畏生命、敬畏使命、敬畏职责。

本书特点如下：

（1）在内容安排上，注重理论知识的全面性，并通过案例分析将理论与飞行实践紧密结合在一起。

（2）在听力题目的编写上，设置灵活，形式多样，互动性强，较好的调动学生的积极性和主动性，确保课堂活动有效开展。

（3）在题目难度梯度设置上，由浅入深，使学生在学习过程中能循序渐进，不断提高飞行专业英语听力水平。

（4）借鉴国外多所航校地面理论课培训内容和方式，开阔学生视野，助力学生较好适应国外的飞行训练。

（5）内容涉及在役经典机型，案例选用与时俱进，为学生从航校到公司搭建了桥梁。

本书涉及的内容较为广泛和专业，编写难度较大。由于编者水平有限，书中不妥之处，希望读者批评指正。

杜丽娟

2020 年 9 月

Contents

Chapter 1   A Brief Introduction to Flying ·· 1

Lesson 1   How Did We Learn to Fly Like the Birds? ·························· 3

Listening 1   Early Days of Flight ··· 3

Listening 2   The Wright Brothers ·· 3

Listening 3   Father of Chinese Aviation ····························· 6

Lesson 2   Aircraft Types ·· 8

Listening 1   Boeing Family ·· 8

Listening 2   Airbus Family ·· 9

Listening 3   COMAC C919 ·· 10

Lesson 3   Aviation Administration and Air Law ····················· 12

Listening 1   Federal Aviation Administration (FAA) ·············· 12

Listening 2   Federal Aviation Regulations (FARs) ··············· 12

Listening 3   1st CAAC EASA Aviation Safety Conference ············ 13

Chapter 2   Aircraft General Knowledge ······························ 15

Lesson 4   Basic Aerodynamics ··· 17

Listening 1   Basic Aerodynamics ·· 17

Listening 2   How Do Aircraft Fly? ······································· 18

Listening 3   Balance ·· 18

Listening 4   Stalls, Stall Recovery and Spins ······················· 19

Lesson 5   Aircraft Structure ·· 22

Listening 1   Cessna 172 ·· 22

Listening 2   Empennage Components ····································· 23

Listening 3   The Powerplant ·· 24

Listening 4   Subcomponents ·· 25

Lesson 6   Flight Instruments Displays ·································· 27

Listening 1   Analog Display: the Six Primary Flight Instruments ············ 27

Listening 2   Digital Display: the Primary Flight Display (PFD) ·········· 28

Listening 3 Magnetic Compass ·· 29

Chapter 3 Pilot Interview and Flight Training ······························· 31

Lesson 7 Pilot Interview ··· 33
Listening 1 Flight School Interview ······························ 33
Listening 2 Airline Company Interview ························· 34
Listening 3 Job Interview Tips ·································· 36

Lesson 8 Flight Training ··· 38
Listening 1 Pilot Study Tips ·································· 38
Listening 2 Private Pilot ····································· 40
Listening 3 Pilot Training Providers ··························· 41

Lesson 9 Pilot Certificates ··· 44
Listening 1 Certificates Types ································· 44
Listening 2 Commercial Pilot Certificate ······················ 45

Chapter 4 Aircraft System ··· 47

Lesson 10 Warning and Fire Protection Systems ····················· 49
Listening 1 Visual, Aural and Tactile Warnings ················· 49
Listening 2 Fire Protection System ····························· 50
Listening 3 South African Airways Flight 295 ················· 51

Lesson 11 Oxygen System ··· 53
Listening 1 Passenger Oxygen ································· 53
Listening 2 Flight Crew Oxygen ······························· 54
Listening 3 Portable Oxygen Cylinders ························· 56

Lesson 12 APU System ··· 58
Listening 1 Definition ·· 58
Listening 2 APU System Description ···························· 58
Listening 3 APU Operation ···································· 60
Listening 4 Korean Airlines Flight 926 ······················· 60

Lesson 13 Fuel System ··· 62
Listening 1 Avoiding a Common Preflight Mistake ··············· 62
Listening 2 Fuel System in Small Aircraft ···················· 62
Listening 3 Fuel System Description ···························· 64
Listening 4 Defueling ·· 65
Listening 5 Avianca Flight 52 ································· 65

Lesson 14   Anti—Icing System ·· 67

Listening 1   Anti—ice System Description ································· 67

Listening 2   Ice Detection System ··· 69

Listening 3   Air Florida Flight 90 ·· 70

Lesson 15   Hydraulic System ·· 71

Listening 1   Hydraulic System Description ······························ 71

Listening 2   Landing Gear System Description ························· 72

Listening 3   Eastern Air Lines Flight 401 ······························ 73

Lesson 16   Communication System ·· 75

Listening 1   Radio Communication System ···························· 75

Listening 2   Communicating ·· 75

Listening 3   Interphone Communication System ····················· 77

Listening 4   Cockpit Voice Recorder ···································· 78

Listening 5   Tenerife Disaster ··· 78

Lesson 17   Flight Control System ·· 80

Listening 1   Flight Controls ·· 80

Listening 2   Primary Flight Controls ···································· 80

Listening 3   Secondary Flight Controls ································· 81

Listening 4   Autopilot ··· 82

Chapter 5   Human Factors ·· 85

Lesson 18   Introduction to Human Factors ····························· 87

Listening 1   What is Human Factors in Aviation? ··················· 87

Listening 2   Visual Illusions ··· 88

Listening 3   Alcohol and Flight ··· 89

Listening 4   Decision Making ··· 89

Appendixes ·· 91

Appendix 1   Keys ··· 91

Appendix 2   Listening Scripts ·· 113

References ·· 172

Chapter 1
A Brief Introduction to Flying

Lesson 1   How Did We Learn to Fly Like the Birds?

Listening 1   Early Days of Flight

1. Matching.

Time	Inventor	Works
In the 19th century.	Daedalus & Icarus.	Drew diagrams of flying machines.
In 17th December, 1903.	Leonardo da Vinci.	Designed the first aircraft – a balloon.
In the 15th century.	The Wright Brothers.	Made over 2, 000 flights in gliders.
In 1783.	The Montgolfier brothers.	Flew their propeller-driven plane.
In ancient Greek.	Otto Lilienthal.	Made wings from feathers.

2. Listen to a story and write the numbers 1~10 next to the words A) to J) as below to show the order in which you hear them.

___A) Airships.

___B) Flying machines.

___C) Gliders.

___D) Accident.

___E) Wings.

___F) Mythology.

___G) Experimenting.

___H) Ballooning.

___I) Lighter-than-air.

___J) Propeller.

Listening 2   The Wright Brothers

1. Listen to the tape, and complete the passage as below.

The Wright Brothers spent (1) _____ in flight. They noticed that birds soared into the wind and that the (2) _____ the curved surface of their wings (3) _____ . Birds change the shape of their wings to (4) _____ _____ .

They believed that they could use this technique to (5) _____ by warping, or changing the shape, of a portion of the wing. Over the next three years, Wilbur and his brother Orville would design a series of (6) _____ which would be flown in (7) _____ flights. They recognized that (8) _____ ____ would be the most crucial and hardest problem to solve. In 1900, the Wrights successfully tested their new 50-pound (9) _____ glider with its 17-foot (10) _____ and wing-warping mechanism at Kitty Hawk, in both unmanned and piloted flights. In fact, it was the first piloted glider. Based upon the results, the Wright Brothers planned to (11) ____ _____ , and build a bigger glider.

2. Listen to the tape, and fill in the table.

The Wright Flyer Notes	
The Wright Flyer (sometimes called the (1) _____ or the (2) _____) was the first (3) _____ , (4) _____ aircraft to fly successfully on (5) _____	
Frame material	(6) _____
Empty weight	(7) _____
Total wingspan	(8) _____
Total length	(9) _____
Total height	(10) _____
Stabilizer position	(11) _____
Rudder position	(12) _____
Engine	
Engine features	(13) _____
Engine position	(14) _____
Propeller	
Function	(15) _____
Propeller material	(16) _____
Propeller length	(17) _____
Horizontal separation	(18) _____
Rotation direction	(19) _____ so as to (20) _____
Cockpit	
You would lie prone in a (21) _____ designed to (22) _____ . A wooden lever in your left hand would (23) _____ and you would have a rudimentary (24) _____ at your	

The Wright Flyer Notes

(25) _____ . A revolution counter was mounted at the base of the engine. All instruments could be turned
off along with the engine by (26) _____ located on the (27) _____

3. Listen to the tape, and answer the questions as below.

(1) What are the problems that occured during they flew the glider in 1901?

_____

_____

_____

| wings | lifting power | forward elevator |
| controlling the pitch | wing−warping mechanism | spin out of control |

(2) How did they solve these problems?

_____

_____

| wind tunnel | wing shapes | glider | 32−foot wingspan |
| tail | stabilize | | |

(3) What are the new elements of the Flyer which weighted 700 pounds?

_____

_____

| motor | accommodate | weight and vibrations |

(4) Why did they build a movable track?

_____

_____

| downhill track | airspeed |

(5) What was the flight record make by the Flyer?

_____

_____

sustained flight	piloted flight

Listening 3   Father of Chinese Aviation

1. Cloze.

On (1) _____ , a biplane with four starting wheels tucked beneath took to the skies in Oakland, California, six years after Orville and Wilbur Wright's first flight. It was the first successful Chinese flight with a self-made plane and the nation's (2) _____ flight. Feng Ru 1 had a wingspan of (3) _____ , a length of (4) _____ and a (5) _____ of 2.45 meters. As the first aircraft designed and constructed by the Chinese, Feng Ru 1's flight marks a leap forward in China's aviation history. On (6) _____ , Feng was killed while performing an (7) _____ before a crowd of 1,000 spectators. He was flying at about 120 feet and had traveled about (8) _____ . His aircraft smashed into a bamboo grove, and his injuries included a pierced lung.

The Republic of China gave Feng Ru a full military funeral, awarding him the posthumous rank of a major general.

At Sun Yat-Sen's request, the words 'Chinese Aviation Pioneer' were engraved upon Feng's tombstone.

(1) A. Sept. 21, 1909   B. Sept. 21, 1919   C. DeC. 21, 1919   D. DeC. 21, 1909

(2) A) famous manned and powered   B) first unmanned and powered

    C) first manned and powered   D) famous unmanned and powered

(3) A) 7.42 meters   B) 6.72 meters   C) 6.32 meters   D) 7.62 meters

(4) A) 7.23 meters   B) 7.43 meters   C) 6.43 meters   D) 6.23 meters

(5) A) high   B) weigh   C) height   D) weight

(6) A) August 26, 1912   B) August 28, 1912

    C) October 26, 1912   D) October 28, 1912

(7) A) aviation exhibition   B) aircraft exhibition

    C) aerial exhibition   D) airshow exhibition

(8) A) five miles before the accident   B) four miles before the accident

    C) fifteen miles before the accident   D) fourteen miles before the accident

Key terms

aircraft	pitch
airship	propeller-driven plane
airspeed	roll control
coordinate	skid
crash	spin
curved surface	out of control
elevator	steam-powered
glide	wingspan
hydrogen gas	tail
landing gear	stabilize
maneuver	motor
pilot	vibration

扫二维码收听 Lesson 1 的音频数字资源

Lesson 2　Aircraft Types

Listening 1　Boeing Family

1. Complete the missing information.

Boeing Family

Product	Size	Seating capacity	Engine number	Maximum range
Boeing 737	midsize, narrow-body	(1) _____	twin-engine	(2) _____
Boeing 747	(3) _____	around 400	(4) _____	up to 7,670 nautical miles
Boeing 757	mid-size, narrow-body	(5) _____	twin-engine	(6) _____
Boeing 767	(7) _____	181 to 375	twin-engine	(8) _____
Boeing 777	wide-body	(9) _____	(10) _____	(11) _____
Boeing 787	(12) _____	(13) _____	twin-engine	8,500 nautical miles

2. Write T if the statement agrees with the information, and F if the statement contradicts the information.

(1) The 787's design features lighter-weight construction with the choice of Aluminum as usual. (　　)

(2) For the engines and fasteners, Boeing 787 applies the material of titanium and steel. (　　)

(3) The special design for the external features points to the engine nacelles with noise-reducing serrated edges and the planar wingtips. (　　)

(4) The most notable contribution to efficiency is applying electrically powered compressors and pumps. (　　)

(5) Another ideal design is to mix the wing ice protection system that uses electro-thermal heater mats and the hot bleed air together efficiently. (　　)

(6) We have got two head-up displays (HUDs) in the cockpit of Boeing 787. (　　)

(7) Like other Airbus airliners, the 787 will use a side-stick as a new try. (　　)

(8) The airplane's control, navigation, and communication systems are networked with the passenger cabin's in-flight internet systems.   (   )

(9) Passengers could get access into these systems when emergency situation happens.

(   )

(10) In order to prevent data transfer, the design contains the air gaps and firewalls.

(   )

3. Completion.

(1) What is Bush's comment on Air Force One?

_____

_____

(2) What are the two places which could show that someone has occupied the Oval Office?

_____

_____

(3) America has many monuments that 1) _____ , but only one can 2) _____ ideal across 3) _____.

Air Force One has 4) _____.

You always had a tremendous feeling inside you, 5) _____. She is an i-con of 6) _____ and a commanding political tool. I think it in some way e-pitomizes the 7) _____, I would say, 8) _____.

Listening 2   Airbus Family

1. Fill in the blanks.

The advanced Airbus product line incorporates a high degree of commonality between family group members.

The Airbus A300 was to be the first aircraft to be developed, manufactured and marketed by Airbus. By (1) _____ the 'A300' label began to be applied to a proposed seat, (2) __ _____.

The A320 is a (3) _____ range twin-engine subsonic (4) _____ transport aircraft introduced as the first (5) _____ aircraft to the Airbus family. The (6) _____ varies between about (7) _____ and (8) _____ pas-sengers.

The A330 is a third-generation, twin-engine (9) _____ aircraft with typically 335 seats in a two-class arrangement. It offers a range of (10) _____ nautical miles with a full complement of (11) _____.

The A380 is a (12) _____ (VLR), subsonic, civil transport aircraft. The A380 has a full-length (13) _____ fuselage. The two passenger decks are referred to as the main and upper deck. Both decks are connected by (14) _____ ____ and (15) _____. The (16) _____ is located between these two decks. Airbus is in tight competition with Boeing every year for aircraft orders. A380, for example, is designed to be larger than the 747.

2. Complete the aircraft profile.

> **Aircraft profile**

● First flight   time: (1) _____

              place: (2) _____

● Served with: (3) _____

● Two-version description: (4) _____

● Control system: (5) _____

Listening 3   COMAC C919

Multiple choice.

(1) What is the feature of C919 according to the tape?

A. China's first co-developed narrow-body passenger jet.

B. China's first home-built narrow-body passenger jet.

C. China's first co-developed large-body passenger jet.

D. China's first home-built large-body passenger jet.

(2) In two or three years, C919 would go through the_____.

A. airborne test and delivery to China Eastern Airlines.

B. airworthiness application and delivery.

C. airborne application and delivery to China Eastern Airlines.

D. airworthiness certification and delivery.

(3) Which of the following test base is not mentioned?

A. Yanliang, Shanxi province.

B. Dongying, Shandong province.

C. Nanchang, Jiangxi province.

D. Yanliang, Shaanxi province.

（4）Which factor is not mentioned in testing?

A. Security .

B. Fuel capability.

C. Convenient.

D. Economy.

（5）Which statement is correct according to the tape?

A. CR929 long-range wide-body aircraft, which is a new generation home-built aircraft.

B. CR929 would like to select suppliers from China, Russia and other Asia-Pacific countries.

C. CR929 will mainly markets in China, Russia and other South African countries.

D. CR929's competing models include the B787 and A330.

Key terms

airliner	in tandem with
body width	jet
cargo	jetliner
cabin	knot
configuration	privilege
category	subsonic
deck	vehicle
freighter	galley
fuselage	

扫二维码收听 Lesson 2 的音频数字资源

Lesson 3　Aviation Administration and Air Law

Listening 1　Federal Aviation Administration (FAA)

1. Give the definition of FAA.

FAA: _____

_____

2. Fill in the blanks.

The Federal Aviation Administration's major roles include:

Regulating U. S. (1) _____ space transportation;

Regulating air (2) _____ facilities' geometry and flight inspection standards;

Encouraging and developing civil (3) _____, including new aviation technology;

Issuing, suspending, or revoking (4) _____;

Regulating (5) _____ to promote safety, especially through local offices called Flight Standards District Offices;

Developing and operating a system of (6) _____ and navigation for both civil and (7) _____ aircraft;

Researching and developing the National Airspace System and civil aeronautics;

Developing and (8) _____ programs to control aircraft noise and other environmental effects of civil aviation.

Listening 2　Federal Aviation Regulations (FARs)

1. Match the words (1) ~ (10) with the definitions A) to J).

(1) certificate	A. Document, or agreement is legally or officially acceptable.
(2) command	B. A special advantage that is given only to one person or group of people.
(3) valid	
(4) issue	C. Someone's ability to do something.
(5) privilege	D. Official permission to do something, or the document giving this permission.
(6) territory	

E. An official paper stating that you have completed study or passed an examination.

(7) identification   F. Be easy to reach or get into.

(8) authorization   G. Give document or things to people by government officially.

(9) capacity   H. To control something.

(10) accessible   I. Land that is owned or controlled by a particular country, ruler,
   or military force.

   J. Official papers or cards that prove who you are.

2. Give the definition of FAR.

FAR: _____

_____

_____

Listening 3   1st CAAC EASA Aviation Safety Conference

1. Matching.

Authority (   ) Bilateral (   ) Manufacture (   ) Navigation (   ) Separation (   )

(1) make (something) on a large scale using machinery;

(2) the process or activity of accurately ascertaining one's position and planning and following a route;

(3) being moved apart;

(4) having or relating to two sides; affecting both sides;

(5) a person or organization having power or control in a particular, typically political or ad-
ministrative, sphere;

2. Fill in the blanks.

The Executive Director of the (1) _____ (EASA), Mr Patrick Ky and
the Deputy Administrator of the (2) _____ (CAAC), Mr Li Jian opened
the first joint Safety Conference organized by the two administrations. The conference brought
together not only the (3) _____, but also leading CEOs from the European and
Chinese aviation industry, as well as attracting more than 40 members of the (4) _____
____media.

The conference marks a new high in (5) _____between the European Union and
China, which have steadily grown closer in recent years. In the presence of CAAC Administra-

tor and the European Commission Director General for Mobility and Transport, leading members of the aviation community discussed progress on establishing a future (6) _____ _____ (BASA) between China and the European Union.

Currently, there are about (7) _____ in use in China. The total fleet is growing quickly with one aircraft entering the market every day. Every new Airbus aircraft is e-quipped with ADS-B technology and about 1,500 aircraft in the market are manufactured by Airbus.

(8) _____surveillance – broadcast (ADS-B) is a surveillance technology in which an aircraft determines its position via (9) _____and periodically broadcasts it, enabling it to be tracked. The information can be received by air traffic control ground stations as a replacement for secondary radar. It can also be received by other aircraft to provide (10) _____and allow self-separation.

The CAAC (11) _____ has published a roadmap for implementation of ADS-B in all civilian aircraft over the next three years.

Key terms

aeronautic	facility
airspace	frequency
altitude	generic
ATC	identification
authority	IFR
authorization	in accordance with
CAAC	issue
certificate	possession
command	privilege
commercial	
dimension	
FAA	

扫二维码收听 Lesson 3 的音频数字资源

Chapter 2
Aircraft General Knowledge

Lesson 4   Basic Aerodynamics

Listening 1   Basic Aerodynamics

1. Fill in the blanks.

Weight is the force of (1) _____ . It acts in a (2) _____ toward the center of the Earth. Lift is the force that acts at a right angle to the direction of (3) _____ . Lift is created by (4) _____ . Thrust is the force that (5) _____ a flying machine in the direction of motion. (6) _____ produce thrust.

Drag is the force that acts (7) _____ the direction of motion.

Drag is caused by (8) _____ and differences in air pressure.

The four forces is shown in Figure 4−1.

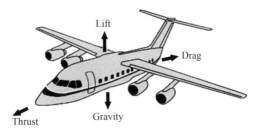

Figure 4−1   The four forces

2. Choose the correct answer.

An aircraft moves in three (1) *dimensions/directions* and is controlled by moving it about one or more of its (2) *axes/axis*. The (3) *longitudinal / lateral* or roll axis extends through the aircraft from nose to tail, with the line passing through the (4) *CG/gravity*. The (5) *longitudinal/lateral* or pitch axis extends across the aircraft on a line through the (6) *wing tips/ winglets*, again passing through the CG. The vertical, or (7) *yaw/roll* axis passes through the aircraft (8) *vertically/eventually*, intersecting the CG. All control movements cause the aircraft to move around one or more of these axes, and allows for the control of the airplane in flight.

3. Complete the picture as below.

The axes of an airplane is shown in Figure 4−2.

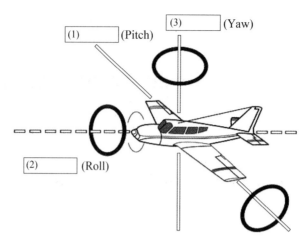

Figure 4-2   Axes of an airplane

Listening 2   How Do Aircraft Fly?

1. Complete the note as below.

(1) Definition of Bernoulli Principle.

_____

_____

(2) What makes an aircraft fly according to the tape?

_____

_____

Listening 3   Balance

1. Choose the correct letter, A), B) or C) .

(1) In steady flight, which statement is true according to what you hear? (   )

A) The lift and thrust are equal.

B) The value of lift and drag are totally the same.

C) The sum of those four forces is zero.

（2）The Newton's Third Law could be applied to prove （　　）

A）The four forces are equal.

B）Each two opposite forces are equal.

C）Each two forces acting on the airplane are equal .

（3）'Thrust equals drag and lift equals weight' is true （　　） .

A）Without exception.

B）When in a straight, level and unaccelerated flight.

C）Except for a straight, level and unaccelerated flight.

（4）In steady flight, which statement is false according to what you hear. （　　）

A）The sum of all upward forces equals the weight.

B）The sum of all forward forces equals the sum of all backward forces.

C）Weight is not equal to the downward force.

2. Choose your answers from the box and write the letters A）to J）in the blanks.

One of the most significant （1）＿＿＿＿＿ of aircraft design is CG. It is the （2）＿＿＿＿＿

point where the mass or weight of an aircraft may be said to center, that is, a point around

which, if the aircraft could be （3）＿＿＿＿＿ or balanced, the aircraft would remain relatively

level. The position of the CG of an aircraft determines the （4）＿＿＿＿＿ of the aircraft in

flight. As the CG moves （5）＿＿＿＿＿ the aircraft becomes more and more dynamically

unstable. In aircraft with （6）＿＿＿＿＿ situated in front of the CG, it is important that the

CG is set with the fuel tank empty. Otherwise, as the fuel is used, the aircraft becomes unsta-

ble. The CG is （7）＿＿＿＿＿＿＿ during （8）＿＿＿＿＿ and construction, and is further

affected by the （9）＿＿＿＿＿of onboard equipment, aircraft （10）＿＿＿＿＿ , and other fac-

tors.

A）suspended	B）rearward	C）loading	D）components	E）installation
F）specific	G）stability	H）fuel tanks	I）initial design	J）computed

Listening 4　Stalls, Stall Recovery and Spins

1. Answer the question as below.

（1）What is the function of MCAS accruing to the tape?

（2）What is the full name of MCAS?

2. Fill in the blanks.

Today I will talk about stalls, stall recovery and spins. An aircraft stall results from a (1) _____ in lift caused by the separation of airflow from the (2) _____ _____ brought on by (3) _____ the critical AOA. A stall can occur at any (4) _____. Stalls are one of the most misunderstood areas of aerodynamics because pilots often believe that (5) _____. In a stall, the wing does not totally stop producing lift. Rather, it can not generate (6) _____. In most straight-wing aircraft, the wing is designed to stall the wing root first. The wing root reaches its (7) _____ first making the stall progress outward toward the wingtip. By having the wing root stall first, aileron effectiveness is maintained at the wingtips, (8) _____ _____. The wing never completely stops producing lift in a stalled condition. If it did, the aircraft would fall to the earth. The stalling speed of a particular aircraft is not a (9) _____ for all flight situations, but a given aircraft always stalls at the same AOA regardless of (10) _____.

3. Write T if the statement agrees with the information, and F if the statement contradicts the information.

(1) Your primary operation when a stall occurs should be contacting the tower about what happened to the aircraft.   (   )

(2) Reducing the angle of attack may result in a secondary stall and / or a spin. (   )

(3) A spin is not a kind of stall which happened when the airplane descends.   (   )

(4) A secondary stall is generally caused by poor stall recovery technique, such as attempting flight before attaining sufficient flying speed.   (   )

4. Answer the questions as below.

What are the normal stall recovery procedures according to the tape?

(1) _____

_____

(2) _____

_____

(3) _____

_____

| AOA | control wheel | recovery | throttle |

Key terms

aerodynamic	lateral
aggravated	longitudinal
airflow	molecule
airspeed	negative pressure
attitude	Newton's Third Law
axes	opposing forces
axis	perpendicular
cargo	positive pressure
center of gravity	powerplant
controllability	pressure differential
corkscrew	propeller
curve	rearward
critical AOA	recovery
Daniel Bernoulli	retard
dimension	slant
disruption	spin
downward	stall
exert	suspend
flightpath	thrust
fluid	vertical
fuselage	wing
gravity	wingtip
guideline	underneath

扫二维码收听 Lesson 4 的音频数字资源

Lesson 5   Aircraft Structure

Listening 1   Cessna 172

1. Fill in the blanks.

The (1) _____ is an American four-seat, (2) _____,
high wing, and (3) _____ aircraft made by the Cessna Aircraft Company. First
flown in (4) _____, more 172s have been built than any other aircraft.

It was developed from (5) _____, using (6) _____,
rather than a tail-dragger (7) _____.

2. Label the illustration of Cessna 172 (shown in Figure 5-1).

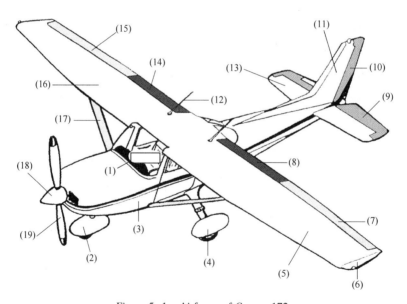

Figure 5-1   Airframe of Cessna 172

Wing tip	radio antenna	left aileron	left flap	elevator
rudder	vertical stabilizer	horizontal stabilizer	right flap	right aileron
right wing	strut	spinner	propeller	nose gear
cockpit	fuselage	main landing gear	left wing	

3.　Answer the questions as below.

（1）What are the basic components of an aircraft?

_____

（2）What is the purpose of the structure?

_____

_____

_____

_____

4.　Write T if the statement agrees with the information, and F if the statement contradicts the information.

（1）Fuselage itself is used to accommodate the crew members, passengers, and cargo.

(　　)

（2）Additionally, Fuselage is designed to make structural connection for the cabin and tail assembly.　　　　　　　　　　　　　　　　　　　　　　　　(　　)

（3）The lifting surfaces in the aircraft are the wings and airfoils.　　(　　)

（4）Aircraft could be designed differently by the manufacturers, but the wings are sharing the same sizes, and shapes.　　　　　　　　　　　　　　　　　　(　　)

（5）Wings may be attached at the top portion of the fuselage, which is called the high wing.

(　　)

（6）Airplanes with a single set of wings are referred to biplanes.　　(　　)

（7）When parked, taxiing, taking off, or landing, the aircraft is supported basically by the landing gear.　　　　　　　　　　　　　　　　　　　　　　　(　　)

（8）Generally speaking, all of the aircrafts are equipped by landing gear with wheels.

(　　)

（9）The powerplant consists of the engine and the propeller.　　(　　)

（10）The engine is designed to provide the power to turn the propeller.　　(　　)

Listening 2　Empennage Components

1.　Fill in the blanks.

The empennage components are shown in Figure 5-2.

2. Complete the table as below.

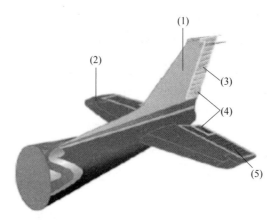

Figure 5-2   Empennage components

Empennage Components

Devices	Components	Location	Functions
Fixed surfaces	vertical stabilizer		
	(2) _____ _____		
(1) _____ _____	(3) _____	the back of the vertical stabilizer	(4) _____ _____
	elevator	(5) _____ _____	to move the (6) _____ up and down during flight
	(7) _____	on the (8) _____ _____	(9) _____

Listening 3   The Powerplant

1. Choose the correct letter, A), B) or C).

(1) Generally, the powerplant consists of (   )

A) Both engines and twin propellers.

B) Engine and propeller.

C) Electrical power and propeller.

(2) Which statement about the engine is true according to the tape? (   )

A) Engine is applied for supplying the power to turn the propeller only.

B) Engine can hardly generate electrical power to provide a vacuum source for some flight instruments.

C) Given most aircraft with only one engine, it could provide source for heating.

(3) The covered housing of the engine is not used to (   )

A) Help cool the engine by ducting air.

B) Assist the cowling, or the nacelle.

C) Streamline the flow of air around the engine.

(4) The function of the propeller is to (   )

A) Provide the thrust.

B) Maintain the balance.

C) Translate the electrical power.

(5) The propeller pulls the airplane forward by (   )

A) The way lift is generated by an airfoil used as a lifting surface or wing.

B) The pressure differential.

C) The electrical power generated by the powerplant.

Listening 4   Subcomponents

1. Listen to a lecture about subcomponents of an airplane, and fill out the outline as below.

The subcomponents of an airplane include the (1) _____ , electrical system,

(2) _____ , and (3) _____ .

    (1) Function of electrical system is to:

    (4) _____ electrical power throughout the airplane;

    (5) _____ , essential systems, such as (6) _____ and

(7) _____ such as (8) _____

    (2) Primary flight controls (9) _____ for (10) _____ , (11) _____ for

(12) _____ , and (13) _____ for (14) _____ , which are operated by

(15) _____ .

    (3) Airplane brakes include (16) _____ whose function is to

(17) _____ . The life of airplane brakes depends on landings rather the mile

because (18) _____ .

Key terms

aerodynamic	landing gear
aileron	monoplane
airfoil	nacelle
airframe	nose

angle of attack	park
anti-icing	payload
assembly	powerplant
attitude	propeller
automatic pilot	roll
biplane	rotor
blade	rudder
brake	stability
cowling	stabilizer
crew	tail
duct	taxi
elevator	thrust
empennage	trailing edge
flight deck	trim tabs
flight instrument	undercarriage
flight path	vacuum
fuselage	vertical stabilizer
horizontal	yaw

扫二维码收听 Lesson 5 的音频数字资源

Lesson 6　Flight Instruments Displays

Listening 1　Analog Display: the Six Primary Flight Instruments

1. Summary.

> **What does analog mean?**

2. Completion (Shown in Figure 6-1).

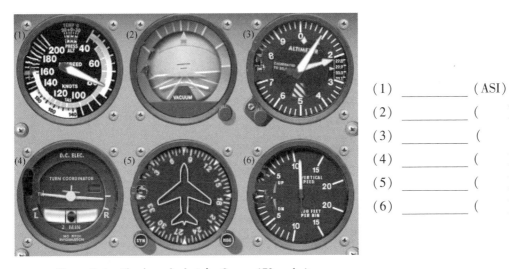

(1) _____ (ASI)

(2) _____ (　　)

(3) _____ (　　)

(4) _____ (　　)

(5) _____ (　　)

(6) _____ (　　)

Figure 6-1　The 'standard six', Cessna 172 cockpit.

3. Fill in the blanks.

Instrument located in the airplane cockpit informs you of the airplane's (1) _____

_____ , (2) _____ , (3) _____ , (4) _____

_____and etc.

Flight instruments are sub-divided into categories according to (5) _____ ;

Pitot-static instruments include instruments indicating (6) _____which op-

erate on (7) _____ ;

Gyroscopic instruments consist of (8) _____ (9) _____

____ ;

Magnetic instrument in the cockpit is the (10) _____ .

4. Fill in the table.

Instrument	Function	Principle
(1) _____ _____	shows the speed at which the aircraft is (2) _____ _____	the pressure exerted by the stream of air (3) _____ _____the local air density, which continually changes as the airplane (4) _____
The Attitude Indicator (AI)	informs the pilot of his or her position in space (5) _____ _____	This is accomplished by (6) _____ the case of the instrument to the aircraft and (7) _____ the displacement of the case with reference to a (8) _____ _____
(9) _____ _____	displays altitude	As the airplane (10) _____ , the relative (11) _____ outside the aircraft changes and the it could report the difference between the (12) ____ _____and a reference, contained in a set of air-tight belows.

Listening 2　Digital Display: the Primary
Flight Display (PFD)

1. Write the complete form of abbreviations.

(1) EFIS _____　　(2) PFD_____

(3) MFD_____　　(4) EICAS_____

(5) SA_____

2. Answer the questions as below.

(1) What is the meaning of the EFIS according to the tape?

_____

(2) What are the components of the EFIS?

_____

（3）What does the PFD indicate?

_____

（4）How does the PFD improve a pilot's situational awareness?

_____

3. Complete the following table.

The Primary Flight Display

Information Provided	Parameters Included	Position on the PFD	Main Components
Airspeed	A）the current（1）_____number B）（2）_____airspeed C）take off and（4）_____reference airspeed and flap（5）_____speeds D）（7）_____and minimum airspeed	A）below the airspeed tape B）（3）_____the airspeed tape C）along the right edge of（6）_____ _____	A）tape B）（10）_____Window C）pointer
Altitude		in a window in the（8）_____of the tape	
Vertical Speed		to the（9）_____of the altitude tape	

Listening 3   Magnetic Compass

1. Complete the summary as below.

Summary

The magnetic compass is considered to be is a reliable source of（1）_____ which requires（2）_____ electrical or suction power. There is a（3）_____ mounted in the magnetic compass.

Variation refers to the（4）_____ between the true and magnetic poles at a given point, with which you could covert（5）_____ to（6）_____. Deviation refers to a（7）_____ which occurs due to disturbances from magnetic fields produced by（8）_____ within the airplane itself. And you will find a chart mounted near the compass which illuminates the（9）_____.

2. Decide if the following statements（1）~（7）are true（T）or false（F）.

（1）Captain Taylor was an experienced pilot.   （   ）

（2）Taylor should have flown west, but he flew east.   （   ）

（3）Taylor believed the air traffic controllers were giving him incorrect information.

（   ）

（4）Taylor could not communicate with the controllers because of bad weather. （   ）

（5）One of the rescue planes made contract with Taylor.   （   ）

（6）People on a ship thought they saw an aircraft crash into the sea.   （   ）

（7）Divers found the missing aircraft of Flight 19 in 1991.   （   ）

Key terms

accessory	magnetic field
air density	multi-function
altimeter	needle
angular	parameter
bar	pitot-static instruments
Bermuda Triangle	pivot
deviation	pole
directional gyro	primary flight display（PFD）
Electronic Flight Instrument System（EFIS）	situational awareness
gyroscope	spring
gyroscopic	suction
indicator	tape
instrument	turn coordinator
in the vicinity of	variation
magnetic compass	

扫二维码收听 Lesson 6 的音频数字资源

Chapter 3
Pilot Interview and Flight Training

Lesson 7   Pilot Interview

Listening 1   Flight School Interview

1. Write down the questions.

(1) _____

(2) _____

(3) _____

(4) _____

(5) _____

(6) _____

(7) _____

(8) _____

(9) _____

(10) _____

2. Oral practice.

Please concern about key words and examples.

What is your greatest strength?

Key words:

Example:

What is your greatest weakness?

Key words:

Example:

3. Fill in the blanks.

Address this question by (1) _____ and (2) _____
_____Focus on (3) _____ rather than personal qualities and bring up
any (4) _____ or other (5) _____ you've used to improve upon your weaknesses. Do not show attempt to (6) _____ either. Interviewers want to see if
you are (7) _____ and that you have the ability to (8) _____ yourself. Don't
(9) _____ a weakness if it is a (10) _____ for the position.

   The interviewer is looking for a strength (11) _____. Be prepared to provide examples of your strength and how you've used it in past study. Some examples of strengths to
highlight include (12) _____, taking the initiative, (13) _____,
working well under pressure, having (14) _____, problem solving, or working well
(15) _____.

Listening 2   Airline Company Interview

   *Airline interviews usually comprise of two parts, a competency assessment and a technical as-
sessment. The technical part of the interview is self explanatory, you could be asked a range of
subjects across the theoretical ATPL spectrum ranging from performance of flight to meteorology.*

   1. Write down the questions.

   (1) _____

   (2) _____

(3) _____

(4) _____

(5) _____

(6) _____

(7) _____

(8) _____

(9) _____

(10) _____

2. Oral practice.

Please concern about key words and examples before your answering.

What qualities make a good pilot?

Key words:

Example:

3. Fill in the blanks.

Employers often seek these qualities in pilots:

Calmness: being able to (1) _____.

Knowledge of (2) _____: pilots know their role in the safety and well being of the employer's customers.

Self motivation: pilots operate separate of (3) _____ and need to not only (4) _____ but other members of the crew.

Teamwork: you must be able to (5) _____ of the role you play within a team outside of piloting duties. A key component of this is (6) _____.

Customer service: a good understanding of (7) _____ is important to an employ-er.

Show an interviewer that you not only meet their (8) _____ but will also be valuable for the company's future or dreams.

Listening 3   Job Interview Tips

1. Completion.

TIP 1. Practice and Prepare

A) (1) _____ the typical job interview questions employers ask ;

B) (2) _____ ;

C) Draw on concrete examples that (3) _____ ____ and back up your resume ;

D) (4) _____during your interview.

TIP 2. Research the Company, Show What You Know

A) Do your homework and research the employer and the industry, so you are ready for the interview question, (5) ' _____ , '

B) if this question is not asked, you should try to (6) _____ what you know about the company on your own.

TIP 3. Try to Stay Calm

A) Try to relax and stay as calm as possible ;

B) Remember that (7) _____ says as much about you as your answers to the questions ;

C) (8) _____ with the interviewer ;

D) Listen to the entire question before you answer ;

E) (9) _____ about your answer.

Key terms

emergency	sop
flight academy	candidate
minimize	employer
emphasize	self-motivation
highlight	demonstrate
accomplishment	familiarize
check-ride	eye contact
crew resource management	option
checklists	

扫二维码收听 Lesson 7 的音频数字资源

Lesson 8   Flight Training

Listening 1   Pilot Study Tips

1. Fill in the blanks.

What is the first step to becoming a pilot?

Decide （1） _____ . FAA's rules for getting a pilot's （2） _____

____ differ depending on the （3） _____ you fly. You can choose among air-

planes, gyroplanes, （4） _____ .

If you are interested in flying （5） _____ , you don't need a pilot's license.

You should also think about what type of flying you want to do. There are several different

types of pilot's licenses, from student pilot all the way up to （6） _____ . The

information below describes the eligibility, training, experience, and testing requirements for

Student Pilots, （7） _____ .

The ground school training （1） is shown in Figure 8–1.

Figure 8–1   Ground school training （1）

2. Cloze.

You should recognize the （1）_____ of planning a definite study program and following it as closely as possible. Haphazard or disorganized （2）_____ usually result in an unsatisfactory score on the （3）_____.

The ideal study program would be to enroll in a （4）_____ course. This offers the advantages of a （5）_____ as well as （6）_____ and training aids designed for pilot instruction. Many of these schools use （7）_____ or programmed instruction materials to supplement classroom instruction.

If you are unable to attend a ground school, the （8）_____ can be satisfactory, provided you obtain the proper study materials and devote a （9）_____ to study. You should establish realistic periodic goals and, equally important, a （10）_____ for completion. （11）_____ is important because it is too easy to 'put off' the study period for some other activity.

A）facilities	B）formal ground school	C）professional instructor
D）audiovisual aids	E）Self-discipline	F）study habits
G）reasonable amount of time	H）knowledge test	I）advantages
J）self-study method	K）target date	

3. Complete the chart.

How to choose a training provider?

Step 1	Step 2	Step 3
Decide a （1）_____	Make a checklist by A）talking to pilots B）（2）_____ Consideration: A）recreational, （3）_____ or professional certificate B）（4）_____ Before your decision making, （5）_____ Do not make your decision only based on （6）_____	（7）_____ different options: A）Evaluate the items on the checklist B）（8）_____ _____

Listening 2   Private Pilot

1. Write T if the statement is true, and F if the statement is false.

(1) A private pilot certificate is unnecessary for the pilots who have higher aviation goals.

(   )

(2) Many private pilots are fond of the fixed one type of private training aircraft.   (   )

(3) Other pilots apply for private certificate for shorten their travel time.   (   )

(4) A private pilot certificate is the most popular one among active pilots in the USA.

(   )

(5) A private pilot certificate gives almost unlimited authority to fly both IFR and VFR.

(   )

(6) A private pilot may carry passengers and fly to earn money.   (   )

(7) Passengers should pay a pro-rata share of flight expenses, such as fuel or rental costs.

(   )

2. Fill in the blanks.

What can you do as a Private Pilot?

With a private pilot certificate, a pilot can (1) _____ from American Flight Schools and fly anywhere in the US. Our (2) _____ enables renters to (3) _____ _____ of aircraft available in the US market. A private pilot certificate is the (4) _____ _____ for adding other ratings and more advanced certificates. American Flight Schools offers most of the (5) _____ and ratings. Popular advancements include: technically advanced (6) _____ aircraft, (7) _____ aircraft, (8) _____ , mountain flying (9) _____ , and (10) _____ .

3. Answer the questions as below.

(1) What are the two main components of the private pilot training?

_____

(2) What are the topics covered during ground school?

_____

(3) How can a student pilot get a private pilot certificate?

_____

(4) What are the common trainer aircraft in private pilot training?

_____

(5) How long does it take?

_____

The ground school training （2）is shown in Figure 8-2.

Figure 8-2   Ground school training （2）

Listening 3   Pilot Training Providers

1. Complete the table.

Differences between an approved pilot school and other training providers

An approved pilot school	Other training providers
Be able to provide a greater variety of （1）_____, dedicated facilities, and more （2）_____.	
Certificated in accordance with Title 14, Code of （3）_____.	Found to be （5）_____ to qualify for FAA certification
Ensures a high quality of training, which meet prescribed standards with respect to （4）_____, and curricula.	
（6）_____ are required to be eligible for a （7）_____.	Longer flight hours

2. Matching and completion.

The example is shown in Figure 8-3.

(a)

(b)

(c)

Figure 8-3   Example

(1) Piper Archer: _____

Fleet number: (4) _____

equipped with: (5) : _____

_____

(2) Cessna Skyhawk 172: _____

Fleet number: (6) _____

equipped with: (7) : _____

_____

(3) Piper Seminole: _____

Fleet number: (8) _____

equipped with: (9) : _____

_____

Key terms

certificate	be compensated for
ultralight vehicles	navigation
eligibility	aerodynamics
Recreational Pilots	aerobatics
Student Pilots	fleet
Private Pilots	training curriculum
haphazard	approved

instructor

audiovisual aids

checklist

evaluate

financial concerns

personnel

curricula

Piper Seminole

Piper Archer

Cessna Skyhawk 172

扫二维码收听 Lesson 8 的音频数字资源

Lesson 9　Pilot Certificates

Listening 1　Certificates Types

1. Answer the questions.

(1) List the pilot certificates.

Type 1 _____

Type 2 _____

Type 3 _____

(2) Fill in the blanks.

For Type 1, experience requirements include at least 1) _____,
including 2) _____ of flight with an instructor and 3) _____
_____.

For Type 2, it does not allow a pilot to fly in 4) _____ (IMC), and the
pilots without an 5) _____ are restricted to daytime flight within 6) _____
_____ (NM) when flying for hire. A complex aircraft must have 7) _____
_____.

For Type 3, it is a prerequisite for acting as a 8) _____ (PIC) of scheduled airline operations. The minimum pilot experience is 9) _____ of flight
time. In addition, the pilot must be at least 23 years of age, be able to read, write, speak,
and understand the English language, and be "of good moral standing."

2. Get additional knowledge about pilot certificates and fill in the table.

Types	Privileges	Limitations	Notes
A) (1) _____ _____	fly in light aircraft within (7) _____ in their local area	just one passenger; (8) _____; No above 10,000 feet; No flying in (9) _____ _____	can be earned in airplane, (10) _____ _____, glider, rotorcraft and lighter-than-air, requires (11) _____ _____ of logged training time

Types	Privileges	Limitations	Notes
B) (2) _____ _____	fly heavier aircraft	NO flights less than (12) _____ from their departure airport; No night flight; NO in controlled airports	requires at least (13) _____ of logged flight time
C) (3) _____ _____	could fly at night and at (14) _____ by any airplane in the category allowed.	No for (15) _____; Not be compensated for pilot services	(16) _____ and at least 40 hours of flight time, 20 of which must be with an instructor
D) (4) _____ _____	allows a pilot to be paid for their flying services	separate regulations for scheduled flights,	fly complex aircraft, which have (17) _____, flaps, and a (18) _____.
E) (5) _____ _____	share their knowledge of flight		
F) (6) _____ _____	fly commercial airliners for a living		have logged at least (19) _____, be at least (20) _____old

Listening 2   Commercial Pilot Certificate

Fill in the blanks.

To obtain a commercial certificate in an airplane under (1) _____ a pilot must have:

A) (2) _____, 100 hours of which must be in (3) _____, and 50 must be in airplanes.

B) 100 hours of (4) _____ time, 50 of which must be in airplanes.

C) 50 hours of (5) _____time, 10 of which must be in an airplane.

D) 20 hours of training, including 10 of (6) _____, 10 of (7) _____, and a smattering of cross-country and (8) _____.

E) 10 hours of (9) _____, including a smattering of cross-country and night.

Additional class ratings, such as adding a (10) _____to single-engine commercial pilot certificate or adding a single-engine rating to a multiengine commercial pilot certificate, will take additional training in that class.

For pilots who train in an approved (11) _____program, the commercial certificate can be earned with less experience, at a minimum of (12) _____.

Key terms

pilot certificates	recreational pilot
compensation	departure airport
rating	high-performance aircraft
instructor	scheduled flight
airmanship	airline transport pilot
nautical mile	endorsement
commercial airplane pilot	multi-engine
turbine-powered	pilot-in-command
prerequisite	cross-country
restrictive	approved
powered parachute	
rotorcraft	
lighter-than-air	

扫二维码收听 Lesson 9 的音频数字资源

Chapter 4
Aircraft System

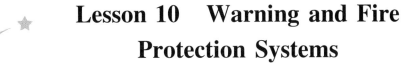

Lesson 10   Warning and Fire Protection Systems

Listening 1   Visual, Aural and Tactile Warnings

1. Complete the table

Large transport aircraft have many (1) _____ used to alert the flight crew to
(2) _____, which include fire and overheat, (3) _____, unsafe landing,
(4) _____, cabin pressure, (5) _____, and (6) _____
systems. Other warning systems used to provide information to the pilots or the flight engineer can be doors, (7) _____, (8) _____, (9) _____ and others.

Condition	Indication
APU fire	(10) _____
malfunction	(11) _____
(12) _____	red lights
wheel well fire	(13) _____
electrical power availability	(14) _____
(15) _____	blue light
(16) _____	red light
(17) _____	green light
landing gear unsafe condition	(18) _____
flight attendant or ground crew communication	(19) _____

2. Complete the outline.

A) The general of warnings.

a) Warn the flight crew to be cautious.

b) The character of the signal is changed according to (1) _____ .

c) (2) _____, (3) _____, and (4) _____ are used singularly or in combinations.

B) Aural warnings

a) Various aural signals call attention to (5) _____ and (6) _____ .

b) A clacker warns for (7) _____ .

c) A warning tone is for (8) _____ .

d) Cabin altitude warns by an intermittent horn.

e) (9) _____ warns by a steady horn.

f) A take-off configuration warns by (10) _____ .

g) Fire warning is by a fire warning bell.

h) Ground proximity warnings and alters are indicated by (11) _____ .

C) (12) _____ .

A stall warning is given at (13) _____ above stall speed because in some configurations the margin between stall and stall warning is less than (14) _____ . We need an artificial stall warning device, (15) _____ which is designed to alert the pilots before (16) _____ develops.

3. Write T if the statement agrees with the information, and F if the statement contradicts the information.

(1) Flight deck panel annunciator lights are used separately from EICAS messages.

(   )

(2) Flight deck panel annunciator lights are used to help locate the affected systems.

(   )

(3) Pilot could get knowledge of potential error by annunciator lights.   (   )

(4) The annunciator lights indicate feedback in response to flight crew action.   (   )

(5) System preflight configuration doesn't connect to the warning light system.   (   )

(6) EICAS is a centrally located crew alerting message display.   (   )

(7) EICAS also displays some system status and maintenance information.   (   )

Listening 2   Fire Protection System

1. Complete the report.

WASHINGTON——The (1) _____ caught fire in Las Vegas nearly three years ago because a (2) _____ cracked from regular usage, federal investigators announced Wednesday.

Nobody died in the incident (3) _____ , but one flight attendant was seriously injured and (4) _____ in the chaotic evacuation, the National Transportation Safety Board found.

After a fire alarm sounded, the captain spent 22 seconds following a (5) _____ before (6) _____ the fuel supply to the (7) _____ , which allowed an estimated 97 gallons of fuel to spill onto the (8) _____ , according to the board.

'If the left engine had been shut down sooner, there would have been (9) _____ to feed the fire', the (10) _____ said in its 30-page report.

2. Complete the table.

Procedure	Operation	Purpose	Condition
Step 1	(1) _____ the warning	to silence (2) _____	
Step 2	A) (3) _____		if (4) _____
	B) move the respective engine (5) _____	Fully closed	
Step 3	(6) _____ the engine start lever	to (7) _____ fuel	
Step 4	A) pull the engine fire warning switch to either (8) _____	to discharge (10) _____	
	B) (9) _____		
	C) (11) _____ the second extinguisher bottle by (12) _____ in the opposite direction	A) to extinguish B) the engine compartment temperature (14) _____	The fire warning switch does not extinguish after (13) _____
Step 5	(15) _____ at the nearest airport		The fire warning light (16) _____ illuminated

3. Dictation.

(1) _____

(2) _____

(3) _____

(4) _____

(5) _____

Listening 3   South African Airways Flight 295

Answer the questions as below.

(1) Please describe the aircraft involved.

(2) What happened to the cargo while the flight was on the way to Mauritius?

(3) What should a captain do in this situation?

(4) What did the flight crew do in dealing with the smoke and fire in the cargo?

(5) What's your opinion on their action? What is the main cause of the fire?

Key terms

amber	intermittent
approach	landing
auto-pilot disconnect	malfunction
aural	margin
buffet	overheat
cabin pressure	overspeed
cancel	proximity
caution	rotate
clacker	simultaneously
compartment	stabilizer
configuration	stick shaker
continuous	switch
device	tactile
discharge	takeoff
extinguish	tone
horn	trim
impending	visual
independent	

扫二维码收听 Lesson 10 的音频数字资源

Lesson 11　Oxygen System

Listening 1　Passenger Oxygen

1. Listen to the tape, and answer the questions.

(1) What happened to JetBlue?

_____

_____

_____

(2) When to deploy the passenger oxygen masks?

_____

_____

_____

2. Fill in the blanks.

Cabin pressurization is a process in which (1) _____ is pumped into the cabin of an aircraft, in order to create a (2) _____ for passengers and crew flying at (3) _____.

For aircraft, this air is usually bled off from the (4) _____ at the compressor stage. The air is cooled, humidified, and mixed with recirculated air if necessary, before it is distributed to the cabin by one or more environmental control systems. The cabin pressure is regulated by the (5) _____. An unplanned drop in the pressure of a (6) _____ system, such as an aircraft cabin, and typically results from (7) _____, material fatigue, (8) _____, or impact, causing a pressure vessel to vent into its (9) _____ surroundings or fail to pressurize at all. Such (10) _____ may be classed as (11) _____.

3. Now you will hear a piece of news. Try to retell the contents by the questions as belows.

(1) What is the main idea of the passage?

(2) Can you describe the aircraft involved in the news?

(3) When and where did the incident happen?

(4) What is the reason for the incident?

(5) How did the aircraft suffer from the incident?

4. Dictation.

(1) _____

(2) _____

(3) _____

(4) _____

(5) _____

5. Choose the correct letter A), B) or C).

(1) A) Different from the flight crew oxygen.   B) Gaseous oxygen.

     C) Chemical oxygen.

(2) A) Twenty minutes   B) Twelve minutes approximately.

     C) Less than twelve minutes.

(3) A) When the oxygen mask doors electrically opened.

     B) When they manually opened.

     C) When they automatically opened.

(4) A) Flight altitude climbs to 14, 000 feet.

     B) Cabin altitude is below 14, 000 feet.

     C) Cabin altitude climbs to 14, 000 feet.

(5) A) Push the passenger oxygen switch to ON. B) In the event of cabin fire.

     C) In the event of cabin depressurization.

Listening 2 Flight Crew Oxygen

1. Write T if the statement agrees with the information, and F if the statement contradicts the information.

(1) The flight crew oxygen system and the passenger oxygen are two separate systems.

                                                    (   )

(2) The flight crew oxygen mask/regulators are located at each crew station. 　(　)

(3) The first observer's oxygen mask/regulator is stored to the right of the first observer's seat. 　(　)

(4) The flight crew use quick-donning diluter demand mask/regulator. 　(　)

(5) For the flight crew system, oxygen is supplied by double cylinders. 　(　)

(6) The crew oxygen pressure can be read out on an indicator. 　(　)

(7) An oxygen cylinder controls the crew oxygen flow and supply low pressure oxygen to a shut off valve. 　(　)

(8) The crew oxygen shut off valve located within easy reach of each crew member.

(　)

(9) Normal flight crew oxygen pressure is 1850 P. S. I. 　(　)

2. Fill in the blanks.

Two (1) _____ oxygen systems are provided, one for the flight crew and one for the passengers. (2) _____ are located throughout the airplane for emergency use.

The flight crew oxygen system uses (3) _____ and (4) _____ located at each crew station.

Oxygen pressure is displayed on the lower (5) _____ . Flight crew and observer masks and regulators are installed in (6) _____ near each seat. (7) _____ the red oxygen mask (8) _____ releases the mask from (9) _____ .

3. Complete the picture as below (shown in Figure 11-1).

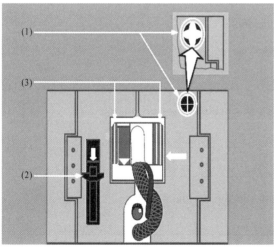

(1) _____
(2) _____
(3) _____

Figure 11-1　Crew oxygen mask panel

4. Fill in the blanks.

Pilot requirements: Use of oxygen

A) Unpressurized aircraft.

Each pilot shall (1) _____ when flying at altitudes above 10,000 feet through (2) _____. For (3) _____ duration; and above 12,000 feet MSL.

B) (4) _____ aircraft. When the (5) _____ more than 10,000 feet MSL, each pilot shall comply with paragraph (A) of this section. When above (6) _____ through 35,000 feet MSL, unless each pilot has an (7) _____ oxygen mask——At least one pilot (8) _____ shall wear an oxygen mask that either supplies oxygen at all times or (9) _____ supplies oxygen whenever the cabin pressure altitude exceeds 12,000 feet MSL; and during that flight, each other pilot shall have an oxygen mask. When above (10) _____, at least one pilot at the controls shall wear an oxygen mask. If one pilot leaves a (11) _____ of an aircraft when operating at altitudes above 25,000 feet MSL, the (12) _____ shall put on oxygen mask until the other pilot returns.

Listening 3   Portable Oxygen Cylinders

1. Answer the questions as below.

Think about:

(1) What is the crucial reason for the accident?

_____

_____

(2) What should a captain do when there is a cabin depressurization?

_____

_____

(3) Who used the portable oxygen cylinder in the accident?

_____

_____

Key terms

afterward	generator
approximately	goggle
baby survival cot	Life jackets
bulkhead	manually
crew station	mask release levers
cylinder	P. S. I.
demo kit	portable
depressurization	quick-donning
fire proof gloves	regulator
first aid kit	slide lever
First Officer	squeeze
flight deck	stowage
forward	stow
torch	

扫二维码收听 Lesson 11 的音频数字资源

Lesson 12   APU System

Listening 1   Definition

1. Fill in the blanks.

The auxiliary power unit (APU) is a (1) _____ located in the airplane
(2) _____ . The APU (3) _____ is located between the
(4) _____ stabilizers on the right side of the airplane. While the primary purpose of the
APU is to supply (5) _____ and (6) _____ on the ground before engine starts,
the APU can also be started (7) _____ . In flight, APU bleed air is available up to
(8) _____ .

2. Label parts of the APU given in the column on the left side (shown in Figure 12-1) .

A. Air Diffuser Duct

B. APU Bleed Air Duct

C. Accessory Cooling Air Duct

D. Air Duct

E. APU Fuel Line

F. Air Inlet Duct

G. Exhaust

(1)

Torque Box
Liner

(2)

(3)

(7)

(4)

Auxiliary
Power Unit
(Shrouded)
(6)

(5)

Bleed Air
Valve

Figure 12-1   Example

Listening 2   APU System Description

1. Dictation.

(1) _____

(2) _____

(3) _____

(4) _____

(5) _____

2. Complete the outline.

(1) APU Start.

A) Electric Start Motor

　　Power Source: (1) _____.

B) Turbine Starter:

　　Power Sources:

　　a) (2) _____.

　　b) (3) _____.

C) Priority to Turbine Starter

　　Condition: (4) _____.

(2) APU Fuel Provided.

A) APU fuel stores in the (5) _____.

B) Pumps operation for APU fuel.

　　a) AC pump: the left forward fuel pump operates (6) _____ under the conditions:

　　　　<1> with AC power available.

　　　　<2> the APU selector in the (7) _____ position.

　　b) DC pump: in the left main tank (8) _____ APU fuel under the conditions:

　　　　<1> AC power (9) _____ available.

　　　　<2> no AC power (10) _____ the left fuel manifold.

(3) APU Starter.

A) Engagement Conditions:

　　a) APU inlet air door in (11) _____.

　　b) at the (12) _____ speed.

　　c) (13) _____ available.

B) Disengagement Condition:

　　a) The APU reaches approximately (14) _____.

Listening 3   APU Operation

1. Complete the table.

Component	Quantity	Light/Indicator/ Switch	Light Color	Operational Condition	Function/Feature
Display Lights	1	APU (1) _____ light	(2) ____	APU oil quantity is (3) _____	for (4) _____ operation
	2	APU (5) _____ light	amber	APU oil pressure is low	a) to initiate an (6) _____ b) to keep (7) _____ till normal pressure
	3	APU (8) _____ light	(9) ____	APU oil temperature is excessive	to initiate an (10) _____
	4	APU (11) _____ light	amber		
Dashboards	1	APU (12) _____ Indicator			with figures from (13) _____
	2	APU (14) _____ AC Ammeter			
Switch	1	(15) _____ switch			with three positions (16) _____

Listening 4   Korean Airlines Flight 926

Think about:

(1) What is the crucial reason for the incident?

_____

_____

(2) What can we learn from this incident?

_____

_____

Key terms

accessory	excessive
ammeter	exhaust
arrow	extended operation
auxiliary	ignition
bleed	initiate
compressor	inlet
consecutive	insufficient
cooling	manifold
dashboard	selector
disengage	self-contained
duct	turbine

扫二维码收听 Lesson 12 的音频数字资源

Lesson 13   Fuel System

Listening 1   Avoiding a Common Preflight Mistake

1. Matching.

(1) rfpeilhtg _____   A) A long strip of ground where airplanes take off and land.

(2) slcekhcit_____   B) A container for holding a liquid or gas.

(3) ayurwn _____   C) Before flight.

(4) nidra _____   D) A list of things to be checked or done.

(5) ktna _____   E) To remove (liquid) from something by letting it flow away or out.

2. Fill in the blanks.

Many years ago, I owned a (1) _____ , and at that time I was based at a little airport in the mountains of California... it was a little (2) _____ with nothing but rocks and trees on both ends of the runways. Certainly there was no chance for a safe landing if you had an (3) _____ right after takeoff. I arrived early one day, just to exercise the airplane, and go for a little scenic flight, (4) _____ . That aircraft had the three (5) _____ , two fuel tanks and a (6) _____ .

I got in the airplane, (7) _____ , and then I pulled out (8) _____ , my pre-flight checklist, that is. A little bit late, but at least I pulled it out. I went through it and I realized I'd forgotten to (9) _____ But fortunately I decided, why not just follow (10) _____ and good practice and do it? I drained 17 cups of water out of that tank.

Listening 2   Fuel System in Small Aircraft

1. Fill in the blanks.

The fuel system is designed to provide (1) _____ of clean fuel from the (2) _____ . The fuel must be available to the engine under all conditions of (3) _____ , and during all approved flight (4) _____ . Two common classifications apply to fuel systems in small aircraft: (5) _____ and fuel-pump systems.

The gravity-feed system utilizes the force of gravity to transfer the fuel from the tanks to the engine. For example, on (6) _____ , the fuel tanks are installed in the wings. This places the fuel tanks above the (7) _____ , and the fuel is gravity fed through the system and into the carburetor.

If the design of the aircraft is such that gravity cannot be used to transfer fuel, (8) _____ are installed. For example, on low-wing airplanes, the fuel tanks in the wings are located (9) _____ . Both gravity-feed and fuel-pump systems may incorporate (10) _____ into the system.

2. Answer the questions as below.

(1) What is the function of a fuel primer?

_____

_____

(2) When should we lock the primer?

_____

_____

(3) What is the function of a strainer?

_____

_____

3. Cloze.

Since these (1) _____ are heavier than aviation fuel, they settle in a sump at the bottom of the (2) _____ assembly. A sump is a low point in a fuel system and/or fuel tank. The fuel system may contain sump, fuel strainer, and fuel tank drains, which may be collocated.

The fuel strainer should be drained before each flight. Fuel samples should be drained and checked (3) _____ for water and contaminants. Water in the sump is (4) _____ . Because in cold weather the water can freeze and block fuel lines. In warm weather, it can flow into the (5) _____ and stop the engine. If water is present in the (6) _____ , more water in the fuel tanks is probable and they should be drained until there is no evidence of water. Never take off until all water and contaminants have been removed from the engine fuel system.

Aircraft with fuel-pump systems have two fuel pumps. The main pump system is (7) _____ with an (8) _____ driven auxiliary pump provided for use in engine starting and in the event the engine pump fails. The (9) _____ pump, also known as a boost pump, provides added reliability to the fuel system. The electrically driven auxiliary pump is controlled by a (10) _____ in the flight deck.

A) visually	B) switch	C) electrically	D) contaminants
E) hazardous	F) sump	G) engine driven	H) auxiliary
I) strainer	J) carburetor	K) flight deck	

Listening 3   Fuel System Description

1. Complete the summary.

(1) Function of the Fuel System.

To supply fuel to (1) _____ and (2) _____ .

(2) Components of the Fuel System.

    A) Tank No. 1.

        Location: integral with the (3) _____ .

    B) (4) _____ .

        Location: (5) _____ with the wing structure.

    C) (6) _____ .

        Location: (7) _____ within the fuselage area.

(3) Tank Pumps of the Fuel System.

    A) Pump quantities: (8) _____ in each fuel tank.

    B) Function:

        a) A single pump: to supply (9) _____ fuel to operate one engine under (10) _____ .

        b) Two center tank pump:

        <1> to have a higher output pressure.

        <2> to (11) _____ the main tank pumps to be used firstly.

        c) Display:

        <1> within more than (12) _____ , the EICAS advisory message (13) _____ is displayed.

        <2> (14) _____ of fuel remaining, (15) _____ EICAS advisory message is displayed.

2. Dictation.

(1) _____

(2) _____

(3) _____

(4) _____

(5) _____

Listening 4　Defueling

1. Fill in the blanks.

Fuel (1) _____, or a fuel (2) _____, is a procedure used by aircraft in (3) _____ before a return to the airport (4) _____, or before landing short of its (5) _____ to (6) _____ aircraft's weight. Aircraft have two major types of weight limits: the (7) _____ and the maximum structural landing weight, with the maximum structural landing weight always being the lower of the two. This allows an aircraft on a (8) _____ to take off at the higher weight, consume fuel en route, and arrive at a lower weight. If a flight takes off at the maximum structural takeoff weight and then faces a situation where it must return to the departure airport (9) _____, there will not be time to (10) _____, and the aircraft may be over the maximum structural landing weight to land back at the departure point.

Listening 5　Avianca Flight 52

1. Answer the questions.

(1) What have you learnt from the talk?

_____

(2) What do you know about the accident they have talked about?

_____

2. Listen to a piece of news and write the numbers 1～10 next to the words A) to J) as below to show the order in which you hear them.

(　　) A) The first officer told the ATC that they could continue to hold for about 5 minutes.

(　　) B) Avianca Flight 52 lost all of its four engines, then it lost its height.

(　　) C) ATC cleared the runway to approach and informed the flight of wind shear.

(　　) D) The crew of Avianca Flight 52 told the ATC that they were running out of fuel.

(　　) E) Avianca Flight 52 had been in a holding pattern near New York due to bad weather.

(　　) F) The pilots of Avianca Flight 52 were forced to abandon the first landing.

（　　）G）The aircraft struck the ground and slid down a hill in the town.

（　　）H）ATC passed Avianca Flight 52 to another controller without mentioning the urgency.

（　　）I）Flight 52 encountered wind shear at an altitude of less than 500 feet.

（　　）J）The crew of Flight 52 thought that they would be landing soon when they were in the holding pattern.

Key terms

abandon	glideslope
alternate	gravity fueling
Avianca	holding pattern
center tank	interconnect
congestion	jettison
crash	JFK
crossfeed valve	manual defueling valve
declare	merely
defuel	override
departure	plunge
due to	pump
dump	receptacle
emergency	single-point pressure fueling station
engine-mounting	sweat
flame out	tank

扫二维码收听 Lesson 13 的音频数字资源

Lesson 14   Anti-Icing System

Listening 1   Anti-ice System Description

1. Answer the questions as below.

(1) Why it's so important to make sure the aircraft is free of snow and ice accumulation before take off?

_____

_____

(2) How to operate the de-ice treatment?

_____

_____

2. Answer the questions as below.

(1) Given Boeing747, what does the anti-ice system include?

_____

_____

(2) Under which conditions will the wing anti-ice be inhibited?

_____

_____

(3) When are the wing anti-ice and nacelle anti-ice generally used?

_____

_____

3. Complete the table.

System	Function	Location
Thermal Anti-icing	to prevent (1) _____ _____ _____	on the (2) _____, engines and (3) _____
Electric Anti-icing	to prevent the formation of ice	on the (4) _____, air data sensor, (5) _____, and potable (6) _____

Continued

System	Function	Location
Ice Detection	to operate (7) _____ , during (8) _____ . automatically	

4. Complete the summary.

Wing Anti-icing System

(1) Function:

To prevent ice on (1) _____ , three, four, (2) _____ , (3) _____ and twelve.

(2) Power: the air from (4) _____ system.

(3) Operational condition:

  A) Airplane (5) _____ .

  B) (6) _____ finds ice.

  C) The flight deck switch is set (7) _____ position.

(4) Component:

  A) A pressure regulating and a (8) _____ .

  B) A pressure sensor inside the leading edge of (9) _____ .

(5) Operational steps:

  A) A perforated spray tube (10) _____ the pressure regulated air (11) _____ the slats.

  B) The air (12) _____ the slats.

  C) The air (13) _____ overboard through (14) _____ in the slat.

Engine Anti-icing System

(1) Function:

To prevent ice on (15) _____ .

(2) Power: (16) _____ air.

(3) Operational condition:

  A) On the (17) _____ or (18) _____ .

  B) The engine anti-icing system must be on when (19) _____ .

(4) Component:

  A) The left valves open when (20) _____ .

  B) The right valves open when the right engine anti-ice switch is placed ON.

  C) (21) _____ provides thirteenth stage bleed air to the nose cowl lips. The

Valves Open Lights illuminate bright while (22) _____ . And then illuminate dim when (23) _____ .

(5) Operation is controlled from the (24) _____ by (25) _____ .

Listening 2   Ice Detection System

1. Choose the correct letter, A), B) or C).

The ice detection system has two (1) _____ , one on each side of the (2) _____ . The ice detection system operates the wing and engine anti-ice system (3) _____ when the airplane is in the air. The detector (4) _____ to the wing and engine anti-ice systems when it finds ice on the detector. This causes the anti-ice systems to come on. Each air data probe has an (5) _____ . The electrical load management system controls the power to each heater element. On the ground with (6) _____ , the system does not heat any probes. On the ground with one or both engines running, these conditions occur. Firstly, total air temperature probes are not heated. Secondly, pitot probes are on (7) _____ . Thirdly, (8) _____ are on full heat. In flight, the system fully heats all probes and sensors.

(1) A) ice sensors   B) ice detectors   C) ice probes

(2) A) forward fuselage   B) forward wing   C) forward

(3) A) always   B) all the tie   C) automatically

(4) A) sends away   B) sends a signal   C) sends a digital

(5) A) electrical heater element B) electric heater elements C) electricity heater elements

(6) A) both engines off   B) both engines on   C) both engines

(7) A) no heat   B) full heat   C) low heat

(8) A) angle of attack sensors   B) angle of attract sensors   C) angle of attack probes

2. Write down the summary.

Listening 3 Air Florida Flight 90

Discuss with your partners about questions as below.

A) How would you describe the weather condition on January 13th, 1982 in Washington, D. C. ?

B) What are the possible causes resulting in the accident in your opinion?

C) How would the ice or snow buildup on the wings degrade the aerodynamic performance?

Key terms

actuate	pitot probe
air data sensor	pneumatic
angle of attack sensor	potable water line
anticipate	predetermined amount
cluster	pressurized disposable con-
dim	tainer
drain mast	probe
droplet	rain repellent
electrical anti-icing	rime
electrically resistive ma-	spray
terial	spread
engine cowl	sublimate
float	Thermal anti-icing
freezing level	total air temperature probe
gage	updraft
hazardous	vapor
lamination	visible moisture
ice detector	visibility
in conjunction with	water repellent coating
perforated spray tube	windshield wiper

扫二维码收听 Lesson 14 的音频数字资源

Lesson 15   Hydraulic System

Listening 1   Hydraulic System Description

1. Answer the questions as below.

(1) Which components of a large airplane are powered by hydraulic system? How about that of the small aircraft?

_____

_____

(2) What are the items that a basic hydraulic system includes for a small airplane?

_____

_____

2. Dictation.

(1) _____

_____

(2) _____

(3) _____

_____

(4) _____

(5) _____

3. Complete the table.

System	Pump	Power Supplier	Function
Left system	A) engine-driven pump	(1) _____	Supply power to do flight controls and (2) _____
	B) AC motor pump	right AC bus	Supply power to the flight controls, (4) _____ And the right thrust reverser
Right system	A) (3) _____ pump		
	B) AC motor pump	Left AC bus	
Center system	A) two AC motor pumps	Left and right (6) _____ buses	Supply power to the flight controls, the leading edge slats, (8) _____ flaps, the alternate reverser, main gear brakes and so on
	B) tow air-driven pumps	A) (7) _____	
	C) a (5) _____ pump	B) auxiliary power unit	

Listening 2　Landing Gear System Description

1. Fill in the blanks.

Given Boeing 747 – 400, there is a two wheel unit (1) _____, two (2) _____, two (3) _____ . Wing gears retraction and extension are held by (4) _____ . Body gears retraction and extension are held by (5) _____ . Nose gears retraction and extension are held by (6) _____ . The gear knows whether it is in the air by using the (7) _____ system by (8) _____ . When the gear is in transit, you can find (9) _____ in the EICAS. When the gear is down and locked, you can find a (10) _____ in the EICAS. When the gear is up and locked, you can find a (11) _____ in the EICAS.

What is the UNIQUE feature of landing gear in Boeing 747?

(12) _____ .

2. Complete the table.

Classifications	Advantage	Disadvantage
Tricycle landing gear airplane	A) allow (1) _____ during landings at high speeds without (2) _____ B) permit better (3) _____ for the pilot during (4) _____ C) tend to prevent (5) _____ by providing more (6) _____ during ground operation	
Tail wheel landing gear airplane		A) (7) _____ of this type aircraft becomes more difficult while on the ground because (8) _____ B) when the tail wheel is (9) _____ , there is a lack of (10) _____

3. Choose the correct letter, A), B), C) or D).

(1) A) Steering.　　B) Braking.　　C) Turning radius.　　D) Absorbing shock.

(2) A) 2.　　　　　B) 14.　　　　　C) 3.　　　　　　D) 6.

(3) A) To improve turning radius.　　B) To improve braking.

　　C) To improve landing.　　D) To improve taxiing.

(4) A) Inboard of each engine.

　　B) Aft of the rear wing spar.

　　C) Below the after bulkhead of the flight compartment.

　　D) In the flight compartment.

4. Complete the summary.

On the ground, the landing gears are locked by automatic lever lock which can be (1) _____ by pushing and holding (2) _____ switch. However, in flight, (3) _____ could automatically (4) _____ the lock. The center hydraulic system is the main power source for retraction, (5) _____, steering and in addition to the (6) _____. The right hydraulic system is the power source for the brakes, but (7) _____ is powered with (8) _____ systems. There is a brake temperature (9) _____ and tire pressure indicator system to show (10) _____ brake temperature and a tire pressure on the Gear Synoptic Display.

Listening 3   Eastern Air Lines Flight 401

Complete the table.

Time of accident	(1) _____ Standard time	Data/Month	(2) _____ _____	Year	(3) _____
Location of accident	(4) _____ miles west northwest of (5) _____	Type of plane	(6) _____ _____	Reason of not landing	System did not indicate (7) _____ _____ in the down position
Population of the passengers	(8) _____	Population of the crew members	(9) _____	Population fatally injured	(10) _____
Probable cause of the accident	a. the failure of the fight crew to (11) _____ during the final (12) _____ b. the failure of the flight crew to (13) _____ soon enough to (14) _____ with the ground				

Key terms

absorbing shock	maintenance
accumulator	nosewheel
braking	relief valve
conventional	reservoir
expel	retraction
extension	simplicity
fatally	steerable
filter	steering
ground looping	synoptic
hydraulic	tailwheel
inherent	tricycle
in tandempar	turning radius
lever lock	unimproved field
mechanism	

扫二维码收听 Lesson 15 的音频数字资源

Lesson 16   Communication System

Listening 1   Radio Communication System

1. Complete the Chart.

```
                    ┌──────────────────────────────────┐
                    │  Radio Communication System      │
                    └──────────────────────────────────┘
```

VHF Communication System

A) frequency range: from

(2)_____to 135 MHz

B) radio components: transmitter,

(3)_____

C) components location: in a cutout

in the (4)_____panel

(5) _____ Component	VHF C Component	VHF R Component

D) operation:

by (6)_____radio tuning

 panel

E) function:

a) to control (7)_____

b) to (8)_____ monitoring

F) configuration:

a) voice or (9)_____

b) normally, VHF C in the (10)_____

HF Communication System

A) frequency range: from 3 MHz

 to (11)_____

B) feature:

Higher power outputs than

 (12)_____

transmitters

HF L	HF R

C) operation:

 by (13)_____

by any radio tuning panel

(1)_____

A) function:

to monitor (14)_____

 VFH radio and two (15)_____

B) operation:

to receive a call from the ground

station the crew is

 (16)_____

Through the communication crew

alerting system

Listening 2   Communicating

1. Fill in the blanks.

The connection between your voice and the radio is via a (1)_____ which contains

the (2)_____ for speaking, and (3)_____ for listening. The headset has two

(4) _____ which are connected to a (5) _____ on the instrument panel or the (6) _____ . All training aeroplanes are required to have headsets and an (7) _____ for communications within the aircraft. There will be a small intercom control panel with two knobs— (8) _____ . The squelch controls the (9) _____ of the microphones. If the squelch is turned down, (10) _____ .

2. Match the situations (1) ~ (7) with the solutions A) ~ G) .

What if the radio doesn't work?

( ) (1) No noise in the headset.

( ) (2) You hear the instructor okay but not yourself.

( ) (3) You have to puff to trigger voice.

( ) (4) Intercom works okay but there is no radio reception.

( ) (5) Still no reception.

( ) (6) The reception is weak.

( ) (7) No reply to your transmissions.

A) Check the microphone plug is fully in.

B) Select the audio selector panel.

C) Check second radio on same frequency.

D) The intercom squelch is set too low.

E) Note if there is a side-tone.

F) Check squelch and volume and try other radio.

G) Check earphone volumes and plug connections, check intercom on and volume and squelch is up.

3. Answer the questions.

What are the important distinctions?

_____

_____

_____

_____

_____

4. Write T if the statement agrees with the information, and F if the statement contradicts the information.

(1) You could start and stop your transmission by control the transmission button. ( )

(2) While you are pressing the transmission button, your silence would be noticed, then other stations could try to call on this particular frequency. ( )

(3) Don't forget to release the PTT at the end of your transmission. ( )

(4) While you're training, you could apply different frequencies as you like. ( )

(5) You can't interrupt any transmissions of lower priority even though there is urgency.

( )

(6) For most communications, we use standard radio phraseology.   (   )

(7) You could transmit with a long silence but you need avoid hesitation sounds.   (   )

(8) Only under some occasion, you could use slang and chat while you're transmitting.

(   )

(9) You could write down what you want to transmit before you do it.   (   )

5. Complete the table.

How to use a microphone correctly?

Dos	Don'ts
1. Actuate the press – to – talk (PTT) switch before commencing to talk	(1)
(2)	(3)
(4)	
(5)	(6)
(7)	
(8)	

Listening 3   Interphone Communication System

1. Fill in the blanks.

The components of Interphone Communication System are:

(1) _____ .

(2) _____ .

(3) _____ .

(4) _____ .

They are normally controlled by (5) _____, and used to (6) _____

_____ .

2. Choose the correct letter, A), B), C) or D) .

(1) A) An independent continuous networks.

    B) An independent communications networks.

    C) An indicated continuous networks.

    D) An indicated communication networks.

(2) A) To provide private communication between cockpit crew members.

    B) To provide private communication between cabin crew members.

    C) To provide private communication between passengers.

    D) To provide private communication between cockpit and cabin.

(3) A) Through a jack.

　　B) Through an interphone.

　　C) Through a wheel.

　　D) Through a PTT switch.

(4) A) Normally.

　　B) Alternately.

　　C) Respectively.

　　D) Directly.

Listening 4　Cockpit Voice Recorder

1. Fill in the blanks.

(1) Definition of CVR:

_____

_____

_____

(2) Definition of FDR:

_____

_____

2. List the function of the audio control panel.

_____

_____

Listening 5　Tenerife Disaster

1. Complete the table.

Data/Month of accident	(1) _____		Year		(2) _____
Location of accident	(3) _____	Type of plane	(4) _____	Weather condition	(5) _____
Original terminal	(6) _____	Reason for diversion	(7) _____	Airliners involved	(8) _____
Cause of the accident	KLM jet was in the (9) _____ and hit the Pan Am plane as it (10) _____ The KLM pilot assumed that he was given (11) _____ due to the ambiguous (12) _____ used at the time				

Key terms

air traffic services（ATS）

annunciate

audio selector panel

boom microphone

Cabin Interphone System

clutter

Cockpit Voice Recorder（CVR）

console

datamode

distress

earphone

flight data recorder（FDR）

Flight Interphone System

hand-held mike

headset

hesitation

HF

hiss

intercom

knob

Mayday

MHz

microphone

pan-pan

Passenger Address（PA）System

Pilot Call Panel（PCP）

plug

press-to-talk（PTT）

radio phraseology

receiver

receptacle

reception

satcom

selective calling system

sensitivity

Service Interphone System

side-tone

simultaneously

socket

squelch

transmission

transmit button

扫二维码收听 Lesson 16 的音频数字资源

Lesson 17　Flight Control System

Listening 1　Flight Controls

1. Fill in the blanks.

Aircraft flight control systems consist of primary and secondary systems.

The primary flight controls are (1) ＿＿＿＿＿. The (2) ＿＿＿＿＿ control these flight control surfaces. The primary flight controls are powered by redundant (3) ＿＿＿＿＿.

Secondary flight controls include a (4) ＿＿＿＿＿＿＿＿. Spoilers operate deferentially to assist (5) ＿＿＿＿＿ for roll control and symmetrically as speed brakes.

2. Answer the questions.

(1) Which component (s) provide the pitch control?

_____

(2) Which component (s) provide the roll control?

_____

(3) Which component (s) provide the yaw control?

_____

(4) What's the function of symmetric spoilers, flaps and slats respectively?

_____

Listening 2　Primary Flight Controls

1. Choose the correct letter, A) , B) , C) or D) .

(1) A) The function of the flight control system.

　　　B) The brief introduction of the flight control system.

　　　C) The location of the flight control system.

　　　D) The controls and indicators of the flight control system.

(2) A) Cables.　　　　B) Pulleys.　　　　C) Pedals.　　　　D) Linkage.

(3) A) Primary flight control system.　　　B) Secondary flight control system.

　　　C) Trim tabs and wing flaps.　　　　D) Elevator, aileron, and rudder.

(4) A) To decrease the speed.　　　　B) To trim the control pressure.

C) To change the wing lifting.　　　D) To connect the movable control surfaces.

2. Complete the summary.

Primary flight control system operation.

A) Controls:

(1) _____ and (2) _____ are used to command control surface.

B) Functions:

a) From the system, (3) _____ and (4) _____ are provided.

b) From the system electronic components, (5) _____ and (6) _____ are provided.

Primary flight control system design.

A) Redundant design:

There are (7) _____ operation modes.

B) Redundant power sources:

a) the primary controls and stabilizer are powered by (8) _____ .

b) flaps and slats are powered with an (9) _____ .

3. Match the components (1)~(3) with the features A)~G).

Components	Features
Ailerons	(1) _____
Elevator	(2) _____
Rudder	(3) _____

A) is connected to the control column in the flight deck by a series of mechanical linkages

B) controls movement of the aircraft about its vertical axis

C) control roll about the longitudinal axis

D) controlled by moving the left or right rudder pedal

E) controls pitch about the lateral axis

F) controlled by moving the control wheel or control stick to the right causes

G) are attached to the outboard trailing edge of each wing and move in the opposite direction from each other

Listening 3　Secondary Flight Controls

1. Dictation.

(1) _____

_____

(2) _____

_____

(3) _____

(4) _____

(5) _____

The A320 flight controls is shown in Figure 17−1.

Figure 17−1   A320 flight controls

Listening 4   Autopilot

1. Fill in the blanks.

To be sure, (1) _____ is widely used in (2) _____ and has been praised for making the skies much safer. But there are growing concerns that the industry is relying too much on automation, especially overly (3) _____ . Boeing, meanwhile, is facing questions about its decisions to with hold information from regulators about the (4) _____ implicated in the Lion Air crash.

A preliminary report from Indonesian investigators indicates that Lion Air 610 crashed because a (5) _____ erroneously reported that the airplane was stalling. The false report triggered an (6) _____ known as (7) _____ System (MCAS) .

Once in flight, the Lion Air crew was unprepared for the automated response set off by the faulty (8) _____ . The pilots fought the automated system, trying to pull the nose back up. They did not succeed.

Ethiopian Airlines 302 was similar to that of the Lion Air 610. Both planes struggled to (9) _____ in the minutes after (10) _____ .

2. Write T if the statement agrees with the information, and F if the statement contradicts the information.

(1) Autopilot is an automatic flight control system that keeps an aircraft among all flight phases.　　　　　　　　　　　　　　　　　　　　　　　　　　　(　)

(2) It can be directed by the pilot, or be coupled to a radio navigation signal.　(　)

(3) Autopilot decreases physical and mental demands on a pilot, which is a hazard to safety.
　　　　　　　　　　　　　　　　　　　　　　　　　　　　　　(　)

(4) Altitude and heading hold are the common features available on an autopilot.　(　)

(5) Gyroscopic attitude indicators and magnetic compasses are applied to control servos connected to the flight control system.　　　　　　　　　　　　　　(　)

(6) The autopilot could be engaged and disengage both automatically and manually.　(　)

(7) In fly-by-wire systems, the autopilot is an integrated component which could not be manually overridden.　　　　　　　　　　　　　　　　　　　(　)

(8) These autopilots work with inertial navigation systems, global positioning systems, and flight computers to control the aircraft.　　　　　　　　　　　(　)

Key terms

autopilot	mechanical linkage
bellcrank	mode
cable	navigation
command	outboard trailing edge
control surface	primary control surface
conventional	pulley
deflection	rearward
disconnect	redundant
fly-by-wire	secondary control surface
gyroscopic	sideward lift
hinge	symmetric
input	tailplane
level flight	trim tab

扫二维码收听 Lesson 17 的音频数字资源

Chapter 5
Human Factors

Lesson 18 Introduction to Human Factors

Listening 1 What is Human Factors in Aviation?

1. Listen to a speech, and answer the questions as below.

(1) What are the two big components in human factors according to his talk?

_____

(2) How dose fatigue influence a pilot in flying?

_____

(3) What is the typical reason for the crash of Colgan Air flight 3407?

_____

(4) What are the sources to a pilot stress?

_____

(5) What is the best way to avoid being stressed?

_____

2. Fill in the blanks.

For years, pilot fatigue has been a real issue. Airline pilots, as well as (1) _____ , can all face fatigue while on the job. Obviously, fatigue is caused by (2) _____ . Here are some specific causes of fatigue: Lack of quality sleep; Sleep disturbances; Interruption of circadian rhythm; (3) _____ (such as family problems, anxiety, or check ride stress); (4) _____ , such as heavy exercise; Poor health, including dehydration or poor diet. Specifically, fatigue in pilots can be caused, or amplified by, the following:

A) Commuting: some pilots start their day 2-3 hours earlier than others to (5) _____ .

B) Layovers at airports: sometimes pilots will have a (6) _____ at an airport, where they are meant to rest.

C) Jet-lag: More apparent with (7) _____ , jet-lag can be a big problem when it comes to pilot fatigue.

D) Night flying: Cargo pilots, especially, deal with fatigue when flying lengthy (8) _____ .

E) due to the imbalance of the body's (9) _____ .

F) Monotonous (10) _____ : Pilots that fly the same aircraft on the same routes into the same airports daily are prone to boredom fatigue.

3. Answer the questions as below.

(1) How do the fatigue and sleep deprivation make influence on the pilots?

_____

_____

(2) Which factors would cause the short−term fatigue?

_____

_____

(3) How to guard against fatigue according to the tape?

_____

_____

Listening 2　Visual Illusions

1. Answer the following questions by deciding if each sentence is true (T) or false (F) .

(1) Seeing is actually a mental process more than a visual process.　　　(　　)

(2) The central foveal region is not effective during the night time.　　　(　　)

(3) Given a wide runway, it will cause an illusion of being too low, and you may take the hold−off too high.　　　(　　)

(4) Since a narrow runway goes with an illusion of being too high, your contact with the runway will be later than expected.　　　(　　)

(5) Your visual judgment of flare height a hold−off prior to touchdown could not changed even if you get knowledge of the runway size ahead, you can allow for this in.　　　(　　)

2. Fill in the blanks.

Most runways are of known (1) _____ and on level ground. On every approach, you should try to achieve the same (2) _____, and your eyes will become accustomed to this, allowing consistent approaches along an (3) _____ merely by keeping your view of the runway through the (4) _____ in a standard perspective.

If approaching a sloping runway, the perspective will be different. A runway that slopes (5) _____will look longer, and you will feel that you are high, when in fact (6) _____. The temptation will be for you to go lower and make a (7) _____ approach. If you know that the runway does have an upslope, (8) _____.

A runway that slopes (9) _____ will kook shorter, and you will feel that you are low, when in fact you are on slope. The tendency will be for you to go higher and make a (10) _____approach. If you know that the runway does have a (11) _____, you can anticipate and avoid this tendency.

Listening 3　Alcohol and Flight

1. Write T if the statement agrees with the information, and F if the statement contradicts the information.

(1) It is the 10th time that the co-pilot in Japan Airlines had been reported drunk before flight.　　　　　　　　　　　　　　　　　　　　　　　　　　　　　　(　)

(2) According to Tokyo's Metropolitan Police, the 42-year-old pilot was found to have 189mg of alcohol per 100ml of blood in his system.　　　　　　　　　　(　)

(3) The legal limit for pilot and drive in the U. K. is both 80 mg.　　　(　)

(4) The co-pilot in Janpan Airlines refused to take a breath test in violation of UK's Law.

(　)

(5) And he was in violation by seeking to fly with a blood alcohol reading equal to the permitted level.　　　　　　　　　　　　　　　　　　　　　　　　　　(　)

(6) The cabin attendant smelled alcohol and called police at once.　　(　)

(7) The drunk co-pilot was scheduled to fly from London Heathrow International Airport to Tokyo just with 244 passengers.　　　　　　　　　　　　　　　　　(　)

(8) The co-pilot admitted to drinking both wine and beer the night before he was scheduled to fly the plane.　　　　　　　　　　　　　　　　　　　　　　　(　)

2. Listen to a lecture, and complete the table as below.

Even small quantities of alcohol in the blood impair a pilot's (1) _____ _____			
A pilot must not fly for at least (2) _____ after drinking the last of any alcohol			
Excessive drinking will lead to (3) _____, or even (4) _____. Too much alcohol can result in (5) _____			
The dangers of alcohol include:	Impairment of (6) _____	Release from (7) _____	Slowed (8) _____	Impaired (9) _____
	Compounding of the (10) _____	Reduced (11) ____ to detail	Increased probability of (12) _____ _____	

Listening 4　Decision Making

1. Fill in the blanks.

Aeronautical decision making, ADM, is a systematic approach to the (1) _____ used by aircraft pilots to consistently determine the (2) _____ in response to a given set of circumstances. (3) _____ is the part of the decision making process which relies on (4) _____, problem recognition, and (5) _____ to reduce risks associated with each flight. Risk Elements in ADM take into consideration the four fundamental risk elements: (6) _____ , and the type of operation that comprise any given aviation situation. It is usually not a single decision that leads to an accident, but a (7) _____ triggered by a number of factors. The poor judgment chain, sometimes referred to as the 'error chain,' is a term used to describe this concept of contributing factors in a human factors–related accident.

Five steps for good decision making are: Identifying (8) _____ to safe flight; Learning behavior modification techniques; Learning how to recognize and (9) _____ ; Developing risk assessment skills; Using all resources in a multicrew situation; Evaluating the (10) _____ .

Key terms

anaemic	flightpath
antibiotics	foveal
anti–depressants	high blood pressure
blackouts	illusion
blood donation	incapacitation
blurred vision	memory loss
chronic	monoxide
compound	prolonged
confusion	retina
copious	rod
deprivation	spasm
Deteriorate	self–perpetuating
distract	sedatives
downslope	shallow
drowsiness	sophisticate
EPT	stimulants
Fatigue	tranquillizers
flare	TUC
	unconsciousness

扫二维码收听 Lesson 18 的音频数字资源

Appendixes

Appendix 1　Keys

Chapter 1　A Brief Introduction to Flying

Lesson 1　How Did We Learn to Fly Like the Birds?

Listening 1　Early Days of Flight

1. Matching.

Time	Inventor	Works
in ancient Greek	Deadalus&Icarus	Made wings from feathers
in the 15th century	Leonardo da Vinci	Drew diagrams of flying machines
in 1783	Montgolfier brothers	Designed the first aircraft—a balloon
in the 19th century	Otto Lilienthal	Made over 2,000 flights in gliders
in 17th December, 1903	Wright Brothers	Flew their propeller−driven plane

2. Listen to a story and write the numbers 1 ~ 10 next to the words A) to J) as below to show the order in which you hear them.

5　3　8　9　2　1　7　4　6　10

Listening 2　The Wright Brothers

1. Listen to tape, and complete the passageas below.

(1) a great deal of time observing birds　(2) flowing over　(3) created lift　(4) turn and maneuver　(5) obtain roll control　(6) gliders　(7) both unmanned (as kites) and piloted　(8) control of the flying aircraft　(9) biplane　(10) wingspan　(11) refine the controls and landing gear

2. Listen to the tape, and fill in the table.　(1) Flyer 1　(2) 1903 Flyer　(3) heavier−than−air　(4) powered　(5) December, 17th, 1903　(6) spruce and ash　(7) 605 ibs　(8) 40 feat, 4 inches　(9) 21 feat, 1 inch　(10) 9 feet, 4 inches　(11) at the front of the aircraft　(12) at the rear　(13) inline, four cylinder water−cooled, aluminum crankcase　(14) to the right of the pilot's cradle　(15) provide both lift and thrust　(16) wooden propellers　(17) 8 feet, 6 inches　(18) 10 feet　(19) left rotated counter clockwise, the right clockwise　(20) reduce negatioe gyroscopic effects on the aircraft in flight　(21) hip cradle　(22) warp

the wings and control the rudder （23）control the elevator （24）instrument panel （25）disposal （26）a single wooden lever （27）lower wing

3. Listen to the tape, and answer the questions as below.

（1）The wings did not have enough lifting power, forward elevator was not effective in controlling the pitch, and the wing-warping mechanism occasionally caused the airplane to spin out of control.

（2）They decided to build a wind tunnel to test a variety of wing shapes and their effect on lift. They planned to design a new glider with a 32-foot wingspan and a tail to help stabilize it.

（3）The Wright Brothers designed a motor and a new aircraft sturdy enough to accommodate the motor's weight and vibrations.

（4）This downhill track would help the aircraft gain enough airspeed to fly.

（5）Orville Wright took the Flyer for a 12-second, sustained flight on December 17th, 1903. This was the first successful, powered, piloted flight in history.

Listening 3   Father of Chinese Aviation

1. Cloze.

（1）A　　（2）C　　（3）D　　（4）B　　（5）C　　（6）A　　（7）C　　（8）A

Lesson 2   Aircraft Types

Listening 1   Boeing Family

1. Complete the missing information.

（1）85 to 215　（2）3.050~5.510　（3）wide body　（4）four wing-mounted engines （5）200 to 289　（6）3,150 to 3,900 nautical miles　（7）mid-size, wide-body　（8）3,850 to 6,385 nautical miles　（9）over 300 passengers　（10）twin-engine　（11）5,235 to 9,380 nautical miles　（12）mid-size, wide-body　（13）210 to 290

2. Write T if the statement agrees with the information, and F if the statement contradicts the information.

（1）F　　（2）F　　（3）F　　（4）T　　（5）F　　（6）T　　（7）F　　（8）T
（9）F　　（10）T

3. Completion.

（1）This is a proud moment for me to walk on the airplane, to remind me of the awesome responsibility of the job.

（2）One is the White House itself; the other's Air Force One.

（3）1）celebrate democracy　2）carry the power of the American　3）the country or around the world　4）has a special aura about it　5）if you were part of history when this airplane flew anywhere　6) speed and power　7) strength and the majesty　8) of the United States of America.

Listening 2　Airbus Family

1. Fill in the blanks.

(1) early 1967　(2) twin engine airliner　(3) short/medium　(4) commercial　(5) single aisle　(6) seating capacity　(7) 120　(8) 179 (9) wide body　(10) 4, 850　(11) passengers and baggage　(12) Very Long Range　(13) double-deck　(14) cabin stairs (15) galley lift　(16) cockpit

2. Complete the aircraft profile.

(1) on 27 April 2005.　(2) Toulouse, France.　(3) Singapore Airlines Emirates, Qantas. (4) the 550-seat passenger version and a freighter.　(5) fully fly-by-wire flight control system.

Listening 3　COMAC C919

Multiple choice.

(1) B　(2) D　(3) D　(4) B　(5) D

Lesson 3　Aviation Administration and Air Law

Listening 1　Federal Aviation Administration (FAA)

1. Give the definition of FAA.

FAA is the national aviation authority of the United States. An agency of the United States Department of Transportation, it has authority to regulate and oversee all aspects of civil aviation in the U. S.

2. Fill in the blanks.

(1) commercial　(2) navigation　　(3) aeronautics　(4) pilot certificates (5) civil aviation (6) air traffic control (7) military　(8) carrying out

Listening 2　Federal Aviation Regulations (FARs)

1. Match the words (1) ~ (10) with the definitions A) to J) .

(1) E　(2) H　(3) A　(4) G　(5) B　(6) I　(7) J　(8) D (9) C　(10) F

2. Give the definition of FAR.

The Federal Aviation Regulations (FARs), which are issued by the FAA, provide rules which apply to all areas of aviation, including flight operations, the construction of aircraft, and the training requirements which must be met to obtain pilot certificates and ratings.

Listening 3　1st CAAC EASA Aviation Safety Conference

1. Matching

Authority (5) Bilateral (4) Manufacture (1) Navigation (2) Separation (3)

2. Fill in the blanks.

(1) European Aviation Safety Agency　(2) Civil Aviation Administration of China　(3) authorities　(4) domestic and international　(5) aviation ties　(6) Bilateral Aviation Safety Agreement　(7) 3,000 civilian aircraft　(8) Automatic dependent　(9) satellite navigation　(10) situational awareness　(11) Flight Standards Department

Chapter 2　Aircraft General Knowledge

Lesson 4　Basic Aerodynamics

Listening 1　Lift and Basic Aerodynamics

1. Fill in the blanks.

(1) gravity　(2) downward direction　(3) motion through the air　(4) differences in air pressure　(5) propels　(6) Engines　(7) opposite to　(8) friction

2. Choose the correct answer.

(1) dimensions　(2) axes　(3) longitudinal　(4) CG　(5) lateral　(6) wing tips　(7) yaw　(8) vertically

3. Complete the picture as below.

(1) lateral axis　(2) longitudinal axis　(3) vertical axis

Listening 2　How Do Aircraft Fly?

1. Complete the note as below.

(1) Fluid pressure decreases as fluid speed increases, and vice versa.

(2) The air going over the top moves faster than the air going underneath, and the air pressure above the wing thus is lower than it is under the wing, where slower moving air molecules bunch together. The pressure differential creates lift.

Listening 3　Balance

1. Choose the correct letter, A), B) or C).

(1) C　(2) B　(3) B　(4) A

2. Choose your answers from the box and write the letters A) to J) in the blanks.

(1) D　(2) F　(3) A　(4) G　(5) B　(6) H　(7) J　(8) I

(9) E　(10) C

Listening 4   Stalls, Stall Recovery and Spins

1. Answer the questions as below.

(1) Automatically prevent stalls under certain conditions.

(2) Maneuvering Charactgeristics Augmentation System.

2. Fill in the blanks.

(1) rapid decrease   (2) wing's surface   (3) exceeding   (4) pitch attitude or airspeed (5) an airfoil stops producing lift when it stalls   (6) adequate lift to sustain level flight   (7) critical AOA   (8) maintaining controllability of the aircraft   (9) fixed value   (10) airspeed, weight, load factor, or density altitude

3. Write T if the statement agrees with the information, and F if the statement contradicts the information.

(1) F   (2) T   (3) F   (4) T

4. Answer the questions as below.

(1) Decrease the angle of attack. Depending on of the type of aircraft, you may find that a different amount of forward pressure on the control wheel is required. Too little forward movement may not be enough to regain lift; too much may impose a negative load on the wing, hindering recovery.

(2) Smoothly apply maximum allowable power. If you are not already at maximum allowable power, increase the throttle to minimize altitude loss and increase airspeed.

(3) Adjust the power as required. As the airplane recovers, you should maintain coordinated flight while adjusting the power to a normal level.

Lesson 5   Aircraft Structure

Listening 1   Cessna 172

1. Fill in the blanks.

(1) Cessna 172 Skyhawk   (2) single-engine   (3) fixed-wing   (4) 1955   (5) the 1948 Cessna 170   (6) tricycle undercarriage   (7) configuration.

2. Label the illustration of Cessna 172.

(1) Cockpit   (2) Nose Gear   (3) Fuselage   (4) Main Landing Gear   (5) Left Wing (6) Wing Tip   (7) Left Aileron   (8) Left Flap   (9) Elevator   (10) Rudder   (11) Vertical Stabllizer   (12) Radio Antenna   (13) Horizontal Stabllizer   (14) Right Flap   (15) Right Aileron   (16) Right Wing   (17) Strut   (18) Splnner   (19) Propeller

3. Answer the questions as below.

(1) Most airplane structures include a fuselage, wings, an empennage, landing gear, and a powerplant.

（2）The purpose of the structure is to transfer the lift from the wings to support the weight of the structure and load, and to protect its payload (passengers and cargo). It also has to carry the thrust from the propeller to pull the aircraft through the air. The tail surfaces are attached to provide stability. The control surfaces allow the pilot to override the stability to change the flight path of the aircraft. On the ground the airframe is supported by the undercarriage.

4. Write T if the statement agrees with the information, and F if the statement contradicts the information.

（1）T   （2）F   （3）F   （4）F   （5）T   （6）F   （7）T   （8）F
（9）T   （10）T

Listening 2   Empennage Components

1. Fill in the blanks.

（1）Vertical stabilizer   （2）horizontal stabilizer   （3）rudder   （4）trim tabs   （5）elevator

2. Complete the table as below.

（1）movable surfaces   （2）horizontal stabilizer   （3）rudder   （4）move the airplane's nose left and right   （5）attached to the back of the horizontal stabilizer   （6）nose of the airplane   （7）trim tabs   （8）ailerons, the rudder, and/or the elevator   （9）reduce control pressures

Listening 3   The Powerplant

1. Choose the correct letter A), B) or C).

（1）B   （2）C   （3）B   （4）A   （5）B

Listening 4   Subcomponents

1. Listen to a lecture about subcomponents of an airplane, and fill out the outline below.

（1）airframe   （2）flight controls   （3）brakes   （4）generate   （5）operate the flight instrument   （6）anti-icing   （7）passenger services   （8）cabin lighting   （9）elevators   （10）pitch   （11）ailerons   （12）roll   （13）rudder   （14）yaw   （15）pilot in flight deck or by an automatic pilot   （16）multiple pads   （17）place pressure on the rotor   （18）airplane brakes are used principally during landings and must absorb enormous amounts of energy

Lesson 6   Flight Instruments Displays

Listening 1   Analog Display: The Six Primary Flight Instruments

1. Summary.

Represectation of actual number.

2. Completion.

(1) airspeed indicator (ASI)　　(2) attitude indicator (AI)

(3) altimeter (ALT)　　　　　　(4) turn coordinator (TC)

(5) directional gyro (DG)　　　(6) vertical speed indicator (VSI)

3. Fill in the blanks.

(1) attitude　(2) direction　(3) altitude　(4) speed　(5) their method of operation　(6) speed, rate of climb or descent, and altitude　(7) air pressure differentials　(8) airplanes' attitude　(9) rate of turn　(10) heading indicator

4. Fill in the table.

(1) The Airspeed Indicator (ASI)　(2) traveling through the air　(3) varies with　(4) climbs or descends　(5) relative to the horizon　(6) fixing　(7) measuring　(8) fixed gyroscope inside　(9) The Altimeter　(10) climbs or descends　(11) air pressure　(12) outside air pressure

Listening 2　Digital Display: the Primary Flight Display (PFD)

1. Write the complete form of abbreviation

1) Electronic Flight Instrument System

2) Primay Flight Display

3) Multi-Function Display

4) Engine Indicating and Crew Alerting System

5) Situational Awareness

2. Answer the questions as below.

(1) An Electronic Flight Instrument System (EFIS) is a flight deck instrument display system in which the display technology used is electronic rather than electromechanical.

(2) EFIS normally consists of a primary flight display (PFD) , multi-function display (MFD) and Engine Indicating and Crew Alerting System (EICAS) display.

(3) The Primary Flight Display (PFD) displays all information critical to flight, including airspeed, altitude, heading, attitude, vertical speed and yaw.

(4) By integrating this information into a single display instead of six different analog instruments, reducing the amount of time necessary to monitor the instruments. PFDs also increase situational awareness by alerting the aircrew to unusual or potentially hazardous conditions

3. Complete the following table.

(1) Mach number　(2) selected　(3) above　(4) landing　(5) maneuvering　(6) airspeed tape　(7) maximum　(8) middle　(9) right　(10) digital

Listening 3 Magnetic Compass

1. Complete the summary as below.

(1) heading information (2) no (3) bar (4) angular difference (5) a true direction
(6) a magnetic direction (7) compass error (8) metals and electrical accessories (9)
remaining error.

2. Decide if the following statements (1) ~ (7) are true T, or false F.

(1) T   (2) F   (3) T   (4) F   (5) F   (6) F   (7) F

Chapter 3 Pilot Interview and Flight Training

Lesson 7 Pilot Interview

Listening 1 Flight School Interview

1. Write down the questions.

(1) Tell me about yourself.

(2) Where do you see yourself in five years?

(3) What is your greatest weakness?

(4) What is your greatest strength?

(5) What qualities make a good pilot?

(6) What was the most difficult decision you've made?

(7) What is a pilot's role during an emergency?

(8) Why do you want to be a pilot?

(9) What have you learnt on campus?

(10) How much do you know about our flight academy?

3. Fill in the blanks.

(1) minimizing your weaknesses (2) emphasizing your strengths (3) professional traits
(4) education (5) opportunities (6) show yourself as perfect (7) self-aware (8)
improve (9) highlight (10) core competency (11) related to the position (12) attention
to detail (13) positive attitude (14) a desire to learn (15) within a team.

Listening 2 Airline Company Interview

1. Write down the questions.

(1) What is your greatest accomplishment?

(2) What have you done recently to improve as a pilot?

(3) What would your leadership style be as a captain?

(4) Have you ever failed a check-ride or required additional training?

(5) What would you do if a crew member had been drinking?

(6) What is the definition of crew resource management (CRM) ?

(7) What is the importance of checklists and SOP?

(8) What makes you qualified for this position?

(9) How are you different from the other candidates?

(10) Why do you want to work for us?

3. Fill in the blanks.

(1) cope in high pressure situations　(2) responsibility　(3) supervisors　(4) motivate themselves　(5) demonstrate awareness　(6) effective communication　(7) customer care (8) current needs

Listening 3　Job Interview Tips

Completion.

(1) review　(2) practice your answers　(3) highlight your skills　(4) listen carefully a list of your own questions　(5) What do you know about this company?　(6) demonstrate　(7) your body language　(8) maintain eye contact　(9) take a moment to think

Lesson 8　Flight Training

Listening 1　Pilot Study Tips

1. Fill in the blanks.

(1) what you want to fly　(2) license (certificate)　(3) type of aircraft　(4) helicopters, gliders, balloons, or airships　(5) ultralight vehicles　(6) airline transport pilot　(7) Recreational Pilots and Private Pilots

2. Cloze.

(1) I　(2) F　(3) H　(4) B　(5) C　(6) A　(7) D　(8) J

(9) G　(10) K　(11) E

3. Complete the chart.

(1) general location　(2) reading articles　(3) private　(4) part-time or full time　(5) visit the provider　(6) financial concerns　(7) evaluate　(8) take some time to think things

Listening 2　Private pilot

1. Write T if the statement is true, and F if the statement is false.

(1) F　(2) F　(3) T　(4) T　(5) F　(6) F　(7) T

2. Fill in the blanks.

(1) rent aircraft　(2) diverse fleet　(3) experience the wide variety　(4) minimum

requirement (5) advanced certificates (6) avionics (7) high performance (8) tail wheel endorsement (9) checkout (10) Cirrus training

3. Answer the questions as below.

(1) Ground school and flight training.

(2) Regulations, airplane systems, navigation, aerodynamics, weather theory, and radio communications.

(3) Certification is achieved upon successful completion of these two tests.

(4) Cessna 172s, Piper Warriors and Archers, Diamond DA40s and DA42s.

(5) Approximately 6 months.

Listening 3　Pilot Training Providers

1. Complete the table.

(1) training aids (2) flexibility in scheduling (3) Federal Regulations part 141 (4) equipment, facilities, personnel (5) impractical (6) fewer flight hours (7) pilot certificate.

2. Matching and completion.

(1) B (2) C (3) A (4) 150 (5) Garmin G500 glass cockpit avionics suites (6) 163 (7) Garmin Moving-Map GPS system (8) 92 (9) a standardized avionics suite

Lesson 9　Pilot Certificates

Listening 1　Certificates Types

1. Answer the questions.

(1) List the pilot certificates.

Type 1: Private Pilot Certificate.

Type 2: Commercial Pilot Certificate.

Type 3: Airline Transport Pilot Certificate.

(2) Fill in the blanks.

1) 40 hours of piloting time 2) 20 hours 3) 10 hours of solo flight 4) instrument meteorological conditions 5) instrument rating 6) 50 nautical miles 7) retractable landing gear, movable flaps, and a controllable pitch propeller 8) pilot in command 9) 1,500 hours

2. Get additional knowledge about pilot certificates and fill in the table.

(1) sport pilot (2) recreational pilot (3) private pilot (4) commercial pilot (5) flight instructor (6) airline transport pilot (7) low altitudes (8) No flying at night (9) Class B, C or D airspace (10) powered parachute (11) 20 hours (12) 50 nautical miles (13) 30 hours (14) controlled airports (15) commercial purposes (16) multiple maneuvers

（17）retractable landing gear （18）controllable-pitch propeller （19）1,500 hours （20）23 years

Listening 2   Commercial Pilot Certificate

1. Fill in the blanks.

（1）FAR Part 61 rules （2）250 hours of flight time （3）powered aircraft （4）pilot-in-command （5）cross-country （6）instrument （7）complex or TAA （8）practical test preparation （9）solo training （10）multi-engine rating （11）FAR Part 141 （12）190 hours

Chapter 4   Aircraft System

Lesson 10   Warning and Fire Protection Systems

Listening 1   Visual, Aural and Tactile Warnings

1. Complete the table.

（1）on-board warning systems （2）unsafe operational conditions （3）unsafe take off （4）overspeed （5）altitude alert （6）ground proximity warning （7）flight controls （8）slat position （9）auto-pilot disconnect （10）red lights （11）amber light （12）engine fire （13）red lights （14）blue lights （15）value position （16）autopilot disconnect （17）landing gear is down and locked （18）red light （19）blue light

2. Complete the outline.

（1）the degree of urgency or hazards involved （2）aural （3）visual （4）tactile （5）warnings （6）cautions （7）airspeed limits （8）autopilot disconnect （9）landing gear positions （10）an intermittent horn （11）voice warnings （12）tactile warnings （13）a minimum of 7% （14）the required 7% （15）a stick shaker （16）a stall

3. Write T if the statement agrees with the information, and F if the statement contradicts the information.

（1）F   （2）T   （3）T   （4）T   （5）F   （6）T   （7）T

Listening 2   Fire Protection System

1. Complete the report.

（1）engine of a British Airways jet （2）compressor （3）Sept. 8, 2015 （4）19 people suffered minor injurie （5）checklist （6）shutting off （7）left engine （8）runway （9）less fuel on the runway （10）board

2. Complete the table.

(1) cancel (2) the fire alarm bell (3) disengage the auto throttle (4) engaged (5) thrust lever (6) move (7) cut off (8) the left or the right (9) hold it for one second (10) a fire extinguisher bottle (11) discharge (12) rotating the switch (13) thirty seconds (14) decreases (15) land (16) remains

3. Dictation.

(1) In the event of APU fire, the APU will automatically shut down.

(2) This action will provide backup for the automatic shutdown feature.

(3) You discharge the extinguishing agent into the APU compartment.

(4) Ensure that the switches are in the normal position.

(5) Test the fire protection system as part of each flight check.

Lesson 11   Oxygen System

Listening 1   Passenger Oxygen

2. Fill in the blanks.

(1) conditioned air (2) safe and comfortable environment (3) high altitudes (4) gas turbine engines (5) outflow valve (6) sealed (7) human error (8) engineering failure (9) lower-pressure (10) decompression (11) Explosive, Rapid, or Slow

3. Try to retell the contents by the questions as below.

(1) Explosive decompression.

(2) The aircraft, with 346 passengers and 19 crew members on board, had an explosive decompression event over the South China Sea while cruising at 29,000 feet on a scheduled flight from Hong Kong China to Melbourne, Australia.

(3) The event happened about 55 minutes into the flight while the aircraft was over the Pacific Ocean, about 200 miles from Manila.

(4) One cylinder associated with the passenger emergency oxygen system, had sustained a sudden failure and forceful discharge of its pressurized contents.

(5) The cylinder had impacted the door frame, door handle and overhead paneling, before presumably falling to the cabin floor and exiting the aircraft through the ruptured fuselage.

4. Dictation.

(1) The passenger's oxygen can be operated both manually and automatically.

(2) On the left side of the cabin there are four oxygen masks to each generator.

(3) It would be necessary to use the passenger oxygen system in the event of cabin depressurization.

(4) The passenger oxygen switch is located on the aft overhead panel.

(5) Passenger oxygen must be used when smoke is generated.

5. Choose the correct letter, A) , B) or C) .

(1) C　(2) B　(3) A　(4) C　(5) C

Listening 2　Flight Crew Oxygen

1. Write T if the statement agrees with the information, and F if the statement contradicts the information.

(1) T　(2) T　(3) T　(4) T　(5) F　(6) T　(7) T　(8) F　(9) T

2. Fill in the blanks.

(1) independent　(2) portable oxygen cylinders　(3) quick－donning masks　(4) regulators　(5) EICAS　(6) oxygen mask panels　(7) squeezing　(8) release lever (9) stowage

3. Complete the picture as below.

(1) Oxygen Flow Indicator　(2) RESET/TEST Switch/ slide lever　(3) Oxygen Mask Release Levers

4. Fill in the blanks.

(1) use oxygen continuously (2) 12, 000 feet MSL (3) 30 minutes (4) Pressurized (5) cabin pressure altitude (6) 25, 000 feet (7) approved quick－donning type (8) at the controls (9) automatically (10) 35, 000 feet MSL (11) pilot duty station (12) remaining pilot at the controls

Lesson 12　APU System

Listening 1　Definition

1. Fill in the blanks.

(1) self－contained gas turbine engine　(2) tail cone　(3) air inlet door　(4) stabilizers (5) electrical power　(6) bleed air　(7) in flight　(8) approximately 17, 000 feet

2. Label parts of the APU given in the column on the left side.

(1) D　(2) C　(3) A　(4) F　(5) B　(6) E　(7) G

Listening 2　APU System Description

1. Dictation.

(1) The APU is started either by an electric start motor or a turbine starter.

(2) On the ground, the electric starter is limited to three consecutive starts.

(3) The APU is a self－contained gas turbine engine.

(4) It contains of two－stage compressor, a turbine and an accessory drive section.

（5）Pressure is sensed in the bleed feeder duct

2. Complete the outline.

（1）APU battery （2）engine bleed air （3）ground cart air （4）sufficient duct pressure （5）left fuel manifold （6）automatically （7）ON （8）provides （9）is not （10）pressurizes （11）the full position （12）proper （13）ignition and fuel （14）50 percent

Listening 3　APU Operation

1. Complete the table.

（1）low oil quantity （2）blue （3）insufficient （4）extended （5）low oil pressure （6）automatic shutdown （7）illuminated （8）high oil temperature （9）amber （10）automatic shutdown （11）overspeed （12）exhaust temperature （13）0～8 （14）generator （15）APU （16）off/on and start

Lesson 13　Fuel System

Listening 1　Avoiding a Common Preflight Mistake

1. Matching.

（1）C　（2）D　（3）A　（4）E　（5）B

2. Fill in the blanks.

（1）Cessna 195 （2）short airport （3）engine problem （4）practice my landings （5）standard fuel drains （6）fuel sump （7）fastened my seatbelt （8）my checklist （9）drain the left tank （10）good discipline

Listening 2　Fuel System in Small Aircraft

1. Fill in the blanks.

（1）an uninterrupted flow （2）fuel tanks to the engine （3）engine power, altitude, attitude （4）maneuvers （5）gravity-feed （6）high-wing airplanes （7）carburetor （8）fuel pumps （9）below the carburetor （10）a fuel primer

2. Answer the questions as below.

（1）Draw fuel from the tanks to vaporize fuel directly into the cylinders prior to starting the engine. （2）When it is not in use. （3）Remove any moisture and other sediments in the system.

3. Cloze.

（1）D　（2）I　（3）A　（4）E　（5）J　（6）F　（7）G　（8）C
（9）H　（10）B

Listening 3 Fuel System Description

1. Complete the summary.

(1) engine (2) APU (3) wing structure (4) Tank No. 2 (5) integral (6) the center tank (7) between the wing roots (8) two (9) sufficient (10) all conditions (11) override (12) 400 pounds of fuel remaining (13) fuel pump center (14) less than 400 pounds (15) fuel low center

2. Dictation.

(1) Engine fuel shutoff valves are located at each engine-mounting wing station.

(2) The engine fuel manifolds are interconnected by use of the crossfeed valve.

(3) Check valves are located throughout the fuel system to ensure the proper direction of fuel flow and to prevent transfer of fuel between tanks.

(4) Center tank check valves open at a lower differential pressure than the check valves in the No. 1 and No. 2 main tanks.

(5) This ensures that center tank fuel is used before main tank fuel, even though all fuel pumps are operating.

Listening 4 Defueling

Fill in the blanks.

(1) dumping (2) jettison (3) certain emergency situations (4) shortly after take off (5) intended destination (6) lighten (7) maximum takeoff weight (8) normal, routine flight (9) due to certain mechanical problems, or a passenger medical issue (10) consume the fuel meant for getting to the original destination

Listening 5 Avianca Flight 52

1. Answer the questions.

(1) Fuel limitation.

(2) The pilots inside the cockpit were talking about their situation. The Captain told the First Officer to declare an emergency. The 'E' word never left the cockpit and the airplane crashed short of the airport, killing numerous people.

2. Listen to a piece of news and write the numbers 1~10 next to the words A) to J) as below to show the order in which you hear them.

(1) E (2) J (3) A (4) H (5) C (6) I (7) F (8) D
(9) B (10) G

Lesson14   Anti-Icing System

Listening 1   Anti- icing System Description

1. Answer the questions as below.

(1) They change the wing's and tail's shape and disrupt the airflow across the surface, hindering the ability to create lift.

(2) Spread the de-icing fluid.

2. Answer the questions as below.

(1) Engine anti-ice (nacelle), wing anti-ice; flight deck window heat; windshield wipers; probe heat.

(2) When wing leading edge flaps extended, or on ground, wing anti-ice is unavailable.

(3) The temperature must be 10℃ or below, visible moisture must be present (eg. clouds, fog, rain, snow or mist, etc.).

3. Complete the table.

(1) the formation of ice (2) wing leading edge (3) engine cowls (4) flight deck windows (5) drain masts (6) water lines (7) these systems (8) icing conditions

4. Complete the summary.

(1) slats two (2) ten (3) eleven (4) pneumatic (5) in the air (6) the ice detection system (7) to the ON (8) shutoff valve (9) each wing (10) carries (11) into (12) heats (13) goes (14) a vent (15) the forward edge of the engine cowl (16) engine bleed (17) ground (18) in flight (19) icing conditions exist or are anticipated (20) the right engine anti-ice switch is placed ON (21) The cowl valve (22) the respective valves are in transit (23) the respective valves are open (24) cockpit (25) individual engine anti-ice switches

Listening 2   Ice Detection System

1. Choose the correct letter, A), B) or C).

(1) B   (2) A   (3) C   (4) B   (5) A   (6) A   (7) C   (8) A

2. Write down the summary.

The formation process and different kinds of icing is included.

Lesson 15   Hydraulic System

Listening 1   Hydraulic System Description

1. Answer the questions below.

(1) Large aircraft: flight control surface, wing flaps, spoilers. Small aircraft: wheel brakes,

retractable landing gear, constant-speed propeller.

(2) Reservoir, pump, filter, selector valve, relief valve, actuator.

2. Dictation.

(1) Hydraulic power is one of the power sources which widely used in the modern airplanes to operate various airplane units and mechanisms.

(2) The hydraulic system of the modern airplanes performs many functions.

(3) Among the units commonly operated by hydraulic systems are landing gear, wing flaps, slats, wheel brakes, nose, wheel steering, and primary flight control surfaces.

(4) The engine-driven pump which is directly coupled to the engine will operate when the engine is running.

(5) Hydraulic systems operate independently at 3,000 Psi normal pressures.

3. Complete the table.

(1) engine (2) left thrust reverser (3) engine-driven (4) the normal main gear brakes (5) ram air turbine (6) AC (7) two engines (8) the trailing edge

Listening 2   Landing Gear System Description

1. Fill in the blanks.

(1) nose gear (2) body gears (3) wing gears (4) hydraulic 2 (5) hydraulic 1 (6) hydraulic 1 (7) air/ground sensing (8) monitoring the tilt of the gear (9) a white cross-hatch (10) green down (11) white up (12) Main gears' steering feature for achieving greater low speed maneuverability

2. Complete the table.

(1) more forceful application of the brakes (2) causing the aircraft to nose over (3) forward visibility (4) takeoff, landing, and taxiing (5) ground looping (6) directional stability (7) directional control (8) the CG located behind the main gear (9) on or near the ground (10) good forward visibility

3. Choose the correct letter, A), B) or C).

(1) C (2) B (3) A (4) C

4. Complete the summary.

(1) manually overridden (2) override (3) air/ground sensing (4) release (5) extension (6) alternative brake system (7) anti-skid protection (8) both (9) monitor system (10) each

Listening 3   Eastern Air Lines Flight 401

1. Complete the table.

(1) 23:42 eastern (2) 29th December (3) 1972 (4) 18 (5) Miami International

(6) Lockheed L-1011 (7) that the nose gear was locked (8) 168 (9) 8 (10) 99 (11) monitor the flight instruments (12) 4 minutes of flight (13) detect an unexpected descent (14) prevent impact

Lesson 16   Communication System

Listening 1   Radio Communication System

1. Complete the Chart.

(1) Selecting Calling (2) 108.0 MHz (3) supply power, and operating controls (4) instrument (5) VHF L (6) any (7) voice transmission (8) receive (9) data (10) data mode (11) 30 MHz (12) VHF (13) tuned (14) three (15) HF radios (16) alerted

Listening 2   Communicating

1. Fill in the blanks.

(1) headset (2) boom microphone (3) earphones (4) plugs (5) socket (6) centre console (7) electronic intercom (8) one for volume and one for squelch (9) sensitivity (10) the background noise is removed

2. Match the situations (1) ~ (7) with the solutions A) ~G) .

(1) G   (2) A   (3) D   (4) B   (5) C   (6) F   (7) E

3. Answer the questions.

(1) The transmit button must be depressed for you to transmit;

(2) The transmitting, most radio sets are unable to receive, and;

(3) Only one transmission from one station within range can occur on the frequency in use without interference. While you are transmitting, no one else can.

4. Write T if the statement agrees with the information, and F if the statement contradicts the information.

(1) T   (2) F   (3) T   (4) F   (5) F   (6) T   (7) F   (8) F   (9) F

5. Complete the table.

(1) Release it until after your message is completed (2) Speak with the microphone close to or just touching your upper lip (3) Significantly vary the distance between your lips and the microphone (4) Speak directly into the microphone (5) Speak a little slower than normal, but at normal volume (6) Raise your voice or shout, or speak as a whisper (7) Pronounce each word clearly and ensure that you clearly annunciate the end of the word (8) Pause briefly before and after the transmission of numbers

Listening 3　Interphone Communication System

1. Fill in the blanks.

(1) Flight Interphone System　(2) Cabin Interphone System　(3) Service Interphone System　(4) Passenger Address System　(5) Audio Control Panel　(6) allow the flight crew to communicate with the flight attendants, ground personnel and maintenance technicians and to make announcements in the passenger cabin

2. Choose the correct letter, A), B), C) or D).

(1) B　(2) A　(3) A　(4) D

Listening 4　Cockpit Voice Recorder

1. Fill in the blanks.

(1) A Cockpit Voice Recorder (CVR), sometimes referred to as a 'black box', is a flight recorder used to record the audio environment in the flight deck of an aircraft for the purpose of investigation of accidents and incidents.

(2) A flight data recorder (FDR) (also ADR, for accident data recorder) is a kind of flight recorder. It is a device used to record specific aircraft performance parameters.

2. The audio control panels are used to manage the radio, satcom, and interphone communication system. Navigation receiver audio can also be monitored.

Listening 5　Tenerife Disaster

1. Complete the table.

(1) March 27　(2) 1977　(3) Los Rodeos　(4) Boeing 747　(5) foggy　(6) Las Palmas　(7) bomb　(8) Panam and KLM　(9) process of taking off　(10) taxied across the runway　(11) clearance to take off　(12) terminology

Lesson 17　Flight Control System

Listening 1　Flight Controls

1. Fill in the blanks.

(1) elevators, ailerons, and rudder　(2) control column　(3) hydraulic systems　(4) a moveable horizontal stabilizer, spoilers, and leading and trailing edge flaps　(5) ailerons

2. Answer the questions.

(1) Two elevators and a movable horizontal stabilizer.

(2) Two ailerons and spoilers and two flaperons.

(3) A single rudder.

(4) Flaps and slats provide high lift for takeoff, approach, and landing. Symmetric spoilers are used as speedbrakes.

Listening 2   Primary Flight Controls

1. Choose the correct letter, A), B), C) or D).

(1) B   (2) C   (3) A   (4) A

2. Complete the summary.

(1) conventional control wheel column (2) pedal inputs (3) conventional control feel (4) pitch response (5) enhanced handling qualities (6) reduce of pilot's workload (7) three (8) hydraulic sources (9) an electrically powered backup system

3. Match the components (1) ~ (3) with the features A) ~ G).

(1) D F G   (2) E A   (3) B D

Listening 3   Secondary Flight Controls

1. Dictation.

(1) Secondary flight control systems may consist of wing flaps, leading edge devices, spoilers, and trim systems.

(2) Then leading edge slats will retract to the extended position when the pitch angle is below the stall attitude.

(3) The Boeing 737 is fitted with six hydraulically powered spoilers on both wings, which are divided into two ground spoilers and four flight spoilers.

(4) The maximum speed for each flap position is presented on a 'flaps limit placard' in the cockpit.

(5) The leading edge devices consist of the leading edge flaps and slats and are both located on the leading side of the wing and are hydraulically powered by hydraulic system B.

Listening 4   Autopilot

1. Fill in the blanks.

(1) automation (2) commercial aviation (3) complex systems (4) anti-stall technology (5) faulty sensor (6) automated system (7) Maneuvering Characteristics Augmentation (8) angle-of-attack data (9) maintain altitude (10) takeoff

2. Write T if the statement agrees with the information, and F if the statement contradicts the information.

(1) F   (2) T   (3) F   (4) T   (5) T   (6) T   (7) F   (8) T

Chapter 5　Human Factors

Lesson 18　Introduction to Human Factors

Listening 1　What is Human Factors in Aviation?

1. Listen to a speech, and answer the questions as below.

(1) Fatigue and stress.

(2) A fatigued pilot will be sluggish and slow to react to an unexpected situation.

(3) Pilots were found to be fatigued and not receiving enough rest.

(4) The weather, their company, the schedule, traffic, etc.

(5) One of the best ways to avoid being stressed is to simply not fly while stressed or if not possible, to always focus on safety when flying.

2. Fill in the blanks.

(1) cargo, corporate and charter pilots　(2) lack of sleep　(3) Mental or emotional stress (4) Physical exertion　(5) commute to work　(6) 12-hour layover　(7) long-haul pilots (8) routes at night　(9) natural circadian rhythm　(10) tasks.

3. Answer the questions as below.

(1) Fatigue, tiredness and sleep deprivation can lower a pilot's mental and physical capacity quite dramatically.

(2) Short-term fatigue can be caused by overwork, mental stress, an uncomfortable body position, a recent lack of sleep, living-it-up a little too much, lack of oxygen or lack of food.

(3) Fly only when your psychological and emotional lives are under control; Maintain a reasonable level of fitness; Eat regularly and sensibly; Have adequate and effective sleep; Ensure that cockpit comfort is optimized and that energy foods and drink are available on long flights; and Exercise your limbs occasionally and, if practicable, land to stretch your legs at least every four hours.

Listening 2　Visual Illusions

1. Answer the following questions by deciding if each sentence is true (T) or false (F).

(1) F　(2) T　(3) F　(4) T　(5) F

2. Fill in the blanks.

(1) width　(2) flight path angle to the horizontal　(3) acceptable approach slope　(4) windscreen　(5) upwards　(6) you are on slope　(7) shallower　(8) you can avoid this tendency　(9) downwards　(10) steeper　(11) downslope

Listening 3　Acohol and Flight

1. Write T if the statement agrees with the information, and F if the statement contradicts the information.

(1) F　(2) F　(3) F　(4) F　(5) F　(6) F　(7) T　(8) T

2. Listen to a lecture, and complete the table as below.

(1) immediate performance and severely affect judgment　(2) 8 hours　(3) confusion, nausea, vomiting and unconsciousness　(4) stop breathing　(5) hangovers, memory loss and blackouts　(6) judgment　(7) inhibitions　(8) reaction time　(9) performance　(10) detrimental effects of hypoxia　(11) attention　(12) being involved in an accident or and incident

Listening 4　Decision Making

1. Fill in the blanks.

(1) mental process　(2) best course of action　(3) Risk Management　(4) situational awareness　(5) good judgment　(6) the pilot, the aircraft, the environment　(7) chain of e-vents　(8) personal attitudes hazardous　(9) cope with stress　(10) effectiveness of one's ADM skills

Appendix 2   Listening Scripts

Chapter 1   A Brief Introduction to Flying

Lesson 1   How Did We Learn to Fly Like the Birds?

Listening 1   Early Days of Flight

1. Matching.

2. Listen to a story and write the number 1~10 next to the words A) to J) as below to show the order in which you hear them.

Listening Script

It is impossible to say when people first dreamed of flying. In ancient Greek mythology Daedalus and Icarus made wings from feathers. In the 15th century Leonardo da Vinci drew diagrams of flying machines. But it wasn't until 1783 that people first flew in an aircraft—a balloon designed by the Montgolfier brothers in France.

The success of the Montgolfiers ensured that ballooning became popular in Europe in the 18th and 19th centuries. This lead to steam-powered airships filled with lighter-than-air hydrogen gas.

The 1800s also saw people experimenting with flying gliders. Englishman Sir George Cayley had success with a glider that was piloted by his coachman. In France, Jean-Marie Le Bris made the first 'higher than departure' flight in 1856. A German, Otto Lilienthal, made over 2,000 flights in gliders before a flying accident killed him in 1896. Not long after, the Wright brothers successfully flew their propeller-driven plane near Ktty Hawk, North Carolina, on December 17th, 1903.

Listening 2   The Wright Brothers

1. Listen to the tape, and complete the sentences as below.

Listening Script

The Wright Brothers spent a great deal of time observing birds in flight. They noticed that birds soared into the wind and that the air flowing over the curved surface of their wings created lift. Birds change the shape of their wings to turn and maneuver. They believed that they could use this technique to obtain roll control by warping, or changing the shape, of a portion of the wing.

Over the next three years, Wilbur and his brother Orville would design a series of gliders which would be flown in both unmanned (as kites) and piloted flights. They recognized that control of the flying aircraft would be the most crucial and hardest problem to solve. In 1900, the Wrights successfully tested their new 50-pound biplane glider with its 17-foot wingspan and wing-warping mechanism at Kitty Hawk, in both unmanned and piloted flights. In fact, it was the first piloted glider. Based upon the results, the Wright Brothers planned to refine the controls and landing gear, and build a bigger glider.

2. Listen to the tape, and fill in the table.

Listening Script

The Wright Flyer (sometimes called the Flyer 1 or the 1903 Flyer) was the first heavier-than-air, powered aircraft to fly successfully. On December 17th, 1903 at 10:35 am, after years of experimentation, the aircraft flew at Kitty Hawk, North Carolina, ushering in the aviation age.

The Wright Flyer was in many ways a natural extension of the gliders that the Wright brothers had built and tested since 1900. However, to account for the extra weight that an engine and propellers would bring to the aircraft, they had to extend the wing area to more than 500 square feet. Extending the wing area of course meant adding even more weight, and by the time of its maiden flight, the Flyer's empty weight reached 605 lbs.

As for the engine, the brothers reached out to many different automobile manufacturers in the hope of finding a lightweight gasoline-powered engine that could efficiently power the aircraft. Since nothing available suited their needs, the Wrights turned to their friend and coworker Charlie Taylor, who was able to build an engine from scratch.

Though the engine design was crude, even by the standards of the day, it had some remarkable features. An inline, four cylinder water-cooled engine, its crankcase was made of aluminum to reduce weight, the first time an aircraft engine had an aluminum component. Today, the majority of aircraft engines are made of aluminum. In operation, the engine could push 12 horsepower.

The engine was mounted to the right of the pilot's cradle, necessitating the extension of the right wing by four inches, bringing the total wingspan of the Flyer to 40 feet, 4 inches. A biplane canard wing design (pitch was controlled by two small stabilizers at the front of the aircraft instead of the more common tail configuration seen on modern airplanes; the rudder was at the rear), the aircraft's total length was 21 feet, 1 inch. Its total height reached 9 feet, 4 inches.

The Wrights realized through experimentation and calculation that a propeller acted as a rotary wing, that it could provide both lift and thrust. They constructed two wooden propellers for the aircraft, each one measuring 8 feet, 6 inches and placed ten feet apart. They were slow turning and rotated away from each other (the left rotated counterclockwise, the right clockwise) so as to reduce negative gyroscopic effects on the aircraft in flight. The propellers were powered by a simple yet effective sprocket and chain transmission, not dissimilar to the operation of a bicycle.

The frame of the Flyer itself was made of both spruce and ash, two types of wood that are lightweight yet durable. The brothers covered the frame of the aircraft in unbleached, untreated muslin to provide a strong yet aerodynamic covering.

If you were to enter the 'cockpit' of the Wright Flyer as the brothers did at Kitty Hawk in the fall of 1903, you would lie prone in a hip cradle designed to warp the wings and control the rudder. A wooden lever in your left hand would control the elevator and you would have a rudimentary instrument panel at your disposal, consisting of a stopwatch and anemometer. A revolution counter was mounted at the base of the engine. All instruments could be turned off along with the engine by a single wooden lever located on the lower wing.

3. Listen to the tape, and answer the questions as below.

Listening Script

In 1901, at Kill Devil Hills, North Carolina, the Wright Brothers flew the largest glider ever flown, with a 22-foot wingspan, a weight of nearly 100 pounds and skids for landing. However, many problems occurred: the wings did not have enough lifting power; forward elevator was not effective in controlling the pitch; and the wing-warping mechanism occasionally caused the airplane to spin out of control. They decided to build a wind tunnel to test a variety of wing shapes and their effect on lift. They planned to design a new glider with a 32-foot wingspan and a tail to help stabilize it.

During 1902, the brothers flew numerous test glides using their new glider. Their studies showed that a movable tail would help balance the craft and the Wright Brothers connected a movable tail to the wing-warping wires to coordinate turns. With successful glides to verify their wind tunnel tests, the inventors planned to build a powered aircraft.

After months of studying how propellers work the Wright Brothers designed a motor and a new aircraft sturdy enough to accommodate the motor's weight and vibrations. The craft weighed 700 pounds and came to be known as the Flyer.

The brothers built a movable track to help launch the Flyer. This downhill track would help the aircraft gain enough airspeed to fly. After two attempts to fly this machine, one of which resulted in a minor crash, Orville Wright took the Flyer for a 12-second, sustained flight on December 17th, 1903. This was the first successful, powered, piloted flight in history.

In 1904, the first flight lasting more than five minutes took place on November 9. The Flyer II was flown by Wilbur Wright.

Listening 3   The first flight

1. Cloze.

Listening Script

Feng Ru 1, the first plane designed and built by the 'Father of Chinese Aviation' Feng Ru, has been restored and donated to the China Science and Technology Museum, to mark the 110th anniversary of China's aviation industry.

On Sept. 21, 1909, a biplane with four starting wheels tucked beneath took to the skies in Oakland, California, six years after Orville and Wilbur Wright's first flight. It was the first successful Chinese flight with a self-made plane and the nation's first manned and powered flight. Feng Ru 1 had a wingspan of 7.62 meters, a length of 7.43 meters and a height of 2.45 meters. As the first aircraft designed and constructed by the Chinese, Feng Ru 1's flight marks a leap forward in China's aviation history, and its restoration is of historical significance.

On August 26th, 1912, Feng was killed while performing an aerial exhibition before a crowd of 1,000 spectators. He was flying at about 120 feet and had traveled about five miles before the accident. His aircraft smashed into a bamboo grove, and his injuries included a pierced lung.

The Republic of China gave Feng Ru a full military funeral, awarding him the posthumous rank of a major general. At SunYat-Sen's request, the words 'Chinese Aviation Pioneer' were engraved upon Feng's tombstone.

Lesson 2   Aircraft Types

Listening 1   Boeing Family

1. Complete the missing information.

Listening Script

The Boeing 737 is a midsize, short-to medium-range, twin-engine narrow-body jet airliner which has developed into a family of nine passenger models with a capacity of 85 to 215 passengers. The 737 is Boeing's only narrow-body airliner in production, with the −600, −700, −800, and −900ER variants currently being built. The Next-Generation 737 is offered in four versions with a seating capacity of 110 to 220 passengers which can fly 3,050~5,510nm.

Design of the Boeing 747, first of the giant jetliners, began in the early 1960s, when market research indicated the need for a much larger capacity transport to cope with the growing passenger and cargo traffic. The 747 is offered in three basic configurations: all passenger, mixed passenger/cargo, and all freighter. The 747−400 is a wide body airplane with four wing-mounted engines and is designed for long range operation with seating capacity of around 400 which can fly non-stop for up to 7,670 nautical miles (14,200km), depending on model.

The Boeing 757 is a mid-size, narrow-body twin-engine jet airliner that was built by

Boeing Commercial Airplanes from 1981 to 2004. The twinjet features a two-crew glass cockpit, turbofan engines, a conventional tail, and for reduced aerodynamic drag, a supercritical wing design. The 757 has a capacity of 200 to 289 persons and a maximum range of 3,150 to 3,900 nautical miles (5,830 to 7,200km), depending on variant.

The Boeing 767 is a mid-size, wide-body twin-engine jet airliner built by Boeing Commercial Airplanes. It was the manufacturer's first wide-body twinjet and its first airliner with a two-crew glass cockpit. The aircraft features two turbofan engines, a conventional tail, and for reduced aerodynamic drag, a supercritical wing design. Designed as a smaller wide-body airliner than preceding aircraft such as the 747, the 767 has a capacity of 181 to 375 persons and a range of 3,850 to 6,385 nautical miles (7,130 to 11,825km), depending on variant.

The Boeing 777 is a long-range, wide-body, twin-engine jet airliner manufactured by Boeing Commercial Airplanes. It is the world's largest twinjet and is commonly referred to as the 'Triple Seven'. The aircraft has seating for over 300 passengers and has a range from 5,235 to 9,380 nautical miles (9,695 to 17,370km), depending on model.

The Boeing 787 Dreamliner is a long-range, mid-size wide-body, twin-engine jet airliner developed by Boeing Commercial Airplanes. It seats 210 to 290 passengers, depending on the variant. The 787-8 and larger 787-9 serve on flights as long as 8,500 nautical miles (15,750km) in three-class seating.

2. Write T if the statement agrees with the information, and F if the statement contradicts the information.

Listening Script

The 787's design features lighter-weight construction. The aircraft is 80% composite by volume. Its materials, listed by weight, are 50% composite, 20% aluminum, 15% titanium, 10% steel, and 5% other. Aluminum is used on wing and tail leading edges, titanium used mainly on engines and fasteners, with steel used in various places. External features include raked wingtips and engine nacelles with noise-reducing serrated edges. The longest-range 787 variant can fly 8,000 to 8,500 nautical miles (15,000 to 15,700km), enough to cover the Los Angeles to Bangkok or New York City to Hong Kong routes. It has a cruising airspeed of Mach 0.85.

Among 787 flight systems, the most notable contribution to efficiency is the new electrical architecture, which replaces bleed air and hydraulic power sources with electrically powered compressors and pumps, as well as completely eliminating pneumatics and hydraulics from some subsystems (e.g., engine starters or brakes). Another new system is a wing ice protection system that uses electro-thermal heater mats on the wing slats instead of hot bleed air that has been traditionally used. An active gust alleviation system, similar to the system used on the

B-2 bomber, improves ride quality during turbulence.

The 787 flight deck includes two head-up displays (HUDs) as a standard feature. Like other Boeing airliners, the 787 will use a yoke instead of a side-stick. The future integration of forward looking infrared into the HUD system for thermal sensing so the pilots can 'see' through the clouds is under consideration.

The airplane's control, navigation, and communication systems are networked with the passenger cabin's in-flight internet systems. Air gaps for the physical separation of the networks, and firewalls for their software separation prevent data transfer from the passenger internet system to the maintenance or navigation systems.

Listening 2　Airbus Family

1. Fill in the blanks.

Listening Script

The advanced Airbus product line incorporates a high degree of commonality between family group members. The Airbus A300 was to be the first aircraft to be developed, manufactured and marketed by Airbus. By early 1967 the 'A300' label began to be applied to a proposed 320 seat, twin engine airliner

The A320 is a short/medium range twin-engine subsonic commercial transport aircraft introduced as the first single aisle aircraft to the Airbus family. The seating capacity varies between about 120 and 179 passengers.

The A330 is a third-generation, twin-engine wide body aircraft with typically 335 seats in a two-class arrangement. It offers a range of 4, 850 nautical miles with a full complement of passengers and baggage.

The A380 is a Very Long Range (VLR), subsonic, civil transport aircraft. The A380 has a full-length double-deck fuselage. The two passenger decks are referred to as the main and upper deck. Both decks are connected by cabin stairs and galley lift. The cockpit is located between these two decks.

Airbus is in tight competition with Boeing every year for aircraft orders. A380, for example, is designed to be larger than the 747.

2. Complete the aircraft profile.

Listening Script

The Airbus A380 is the largest jet airliner ever built and is the world's first full double-deck passenger aircraft. First flight took place from Toulouse, France on 27 April 2005 and led to a flight test program involving five Airbus A380s. Service entry with launch customer Singapore Airlines happened on the 25th October 2007, with A380s also now in service with Emirates and Qantas. Total orders for the Airbus A380 stand at 159, from 16 customers. Two versions of the Airbus A380 are currently offered, the 550-seat passenger version and a freighter, the Airbus A380F, with stretched versions of both planned. As with all modern Airbus types, the Airbus A380 is controlled by a fully fly-by-wire flight control system.

Listening 3 COMAC C919

1. Multiple choice.

Listening Script

Commercial Aircraft Corp of China, the manufacturer of China's first home-built narrow-body passenger jet C919, said it will start manufacturing the first aircraft later this year and it aims to get airworthiness certification and deliver to China Eastern Airlines in two to three years.

Now, four C919 test jets——the 101, 102, 103 and 104 prototypes——are in intense flying tests, static tests and other ground tests at test bases including Yanliang district in Xi'an, Shaanxi province, Dongying in Shandong province, and Nanchang in Jiangxi province. Another two new test jets, the 105 and 106, will be put into tests within this year. So far, 815 orders have been placed for the C919 from home and abroad.

'The manufacturing of the C919 will help the Chinese aviation industry to be involved in the world's supply chain system of large aircraft, and China will be able to acquire valuable experience,' said Wang Yanan, editor-in-chief of Aerospace Knowledge magazine.

'The C919 still needs to undergo several tests to showcase that it is a safe, fuel-efficient, and convenient aircraft. The Chinese aviation industry could transform from a manufacturing giant to an innovation power,' he said.

Meanwhile, the CR929 long-range wide-body aircraft, which is co-developed by China and Russia, has finished conceptual designs, and now it is in the stage of selecting suppliers. The CR929 will mainly target markets in China, Russia and other Asia-Pacific countries. Its competing models include the Airbus'A330 and Boeing's B787.

Lesson 3 Aviation Administration and Air Law

Listening 1 Federal Aviation Administration (FAA)

1. Give the definition of FAA.

Listening Script

The Federal Aviation Administration (FAA) is the national aviation authority of the United States. An agency of the United States Department of Transportation, it has authority to regulate and oversee all aspects of civil aviation in the U.S. The Federal Aviation Act of 1958 created the organization under the name 'Federal Aviation Agency', and adopted its current name in 1966 when it became a part of the United States Department of Transportation.

2. Fill in the blanks.

Listening Script

The FAA's major roles include:

(1) Regulating U. S. commercial space transportation.

(2) Regulating air navigation facilities' geometry and flight inspection standards.

(3) Encouraging and developing civil aeronautics, including new aviation technology.

(4) Issuing, suspending, or revoking pilot certificates.

(5) Regulating civil aviation to promote safety, especially through local offices called Flight Standards District Offices.

(6) Developing and operating a system of air traffic control and navigation for both civil and military aircraft.

(7) Researching and developing the National Airspace System and civil aeronautics.

(8) Developing and carrying out programs to control aircraft noise and other environmental effects of civil aviation.

Listening 2   Federal Aviation Regulations (FARs)

2. Give the definition of FAR.

Listening Script

The Federal Aviation Regulations (FARs), which are issued by the FAA, provide rules which apply to all areas of aviation, including flight operations, the construction of aircraft, and the training requirements which must be met to obtain pilot certificates and ratings. The FAR are identified by a specific title number within the larger group of rules contained in the Code of Federal Regulation (CFR).

The FARs are divided into numbered parts (FAR Part 61, FAR Part 91, etc.) and regulations are typically identified by the part number, followed by the specific regulation number, for example, FAR 91. 106. During your training, you will become familiar with the regulations which apply to you.

Listening 3   1st CAAC EASA Aviation Safety Conference

2. Fill in the blanks.

Listening Script

The Executive Director of the European Aviation Safety Agency (EASA), Mr Patrick Ky and the Deputy Administrator of the Civil Aviation Administration of China (CAAC), Mr Li Jian opened the first joint Safety Conference organized by the two administrations. The conference brought together not only the authorities, but also leading CEOs from the European and Chinese aviation industry, as well as attracting more than 40 members of the domestic and international media.

The conference marks a new high in aviation ties between the European Union and China, which have steadily grown closer in recent years. In the presence of CAAC Administrator and the European Commission Director General for Mobility and Transport, leading members of the aviation community discussed progress on establishing a future Bilateral Aviation Safety Agreement (BASA) between China and the European Union.

Currently, there are about 3,000 civilian aircraft in use in China. The total fleet is growing quickly with one aircraft entering the market every day. Every new Airbus aircraft is equipped with ADS-B technology and about 1,500 aircraft in the market are manufactured by Airbus.

Automatic dependent surveillance——broadcast (ADS-B) is a surveillance technology in which an aircraft determines its position via satellite navigation and periodically broadcasts it, enabling it to be tracked. The information can be received by air traffic control ground stations as a replacement for secondary radar. It can also be received by other aircraft to provide situational awareness and allow self-separation.

The CAAC Flight Standards Department has published a roadmap for implementation of ADS-B in all civilian aircraft over the next three years. According to CAAC, the current number of civilian aircraft equipped with the technology is already above 80%. Major investments in ADS-B ground stations are planned for the same period and European companies such as Thales are ready to support the process of building up and implement this technology.

Chapter 2　Aircraft General Knowledge

Lesson 4　Basic Aerodynamics

Listening 1　Basic Aerodynamics
1. Fill in the blanks.
Listening Script

Weight is the force of gravity. It acts in a downward direction——toward the center of the Earth.

Lift is the force that acts at a right angle to the direction of motion through the air. Lift is created by differences in air pressure.

Thrust is the force that propels a flying machine in the direction of motion. Engines produce thrust.

Drag is the force that acts opposite to the direction of motion. Drag is caused by friction and differences in air pressure.

2. Choose the correct answer.

Listening Script

An aircraft moves in three dimensions and is controlled by moving it about one or more of its axes. The longitudinal or roll axis extends through the aircraft from nose to tail, with the line passing through the CG. The lateral or pitch axis extends across the aircraft on a line through the wing tips, again passing through the CG. The vertical, or yaw, axis passes through the aircraft vertically, intersecting the CG. All control movements cause the aircraft to move around one or more of these axes, and allows for the control of the airplane in flight.

Listening 2　How Do Aircraft Fly?

1. Complete the note as below.

Listening Script

Airplanes fly when the movement of air across their wings creates an upward force on the wings (and thus the rest of the plane) that is greater than the force of gravity pulling the plane toward the earth.

The physics behind this phenomenon was first described by Daniel Bernoulli, an 18th century Swiss mathematician and scientist who studied the movement of fluids. Bernoulli discovered that the pressure exerted by a moving fluid is inversely proportional to the speed of the fluid. In other words, fluid pressure decreases as fluid speed increases, and vice versa.

The same principle applies to moving air. The faster that air moves through a space, the lower the air pressure; The slower it moves, the higher the pressure. Aircraft wings are designed to take advantage of that fact and create the lift force necessary to overcome the weight of the aircraft, and get airplanes off the ground. The undersides of wings are more or less flat, while their tops are curved. In addition, wings are slanted slightly downward from front to back, so air moving around a wing has a longer way to travel over the top than it does underneath. The air going over the top moves faster than the air going underneath, and the air pressure above the wing thus is lower than it is under the wing, where slower moving air molecules bunch together. The pressure differential creates lift, and the faster the wing moves through the air, the greater the lift becomes, eventually overcoming the force of gravity upon the aircraft.

Listening 3　Balance

1. Choose the correct letter, A), B) or C).

Listening Script

Let's talk more about lift, thrust, weight and drag. In steady flight, the sum of these opposing forces is always zero. There can be no unbalanced forces in steady, straight flight based upon Newton's Third Law, which states that for every action or force there is an equal, but opposite, reaction or force. This is true whether flying level or when climbing or descending.

It does not mean the four forces are equal. It means the opposing forces are equal to, and thereby cancel, the effects of each other. The usual explanation states that thrust equals drag and lift equals weight. Although basically true, this statement can be misleading. It should be understood that in straight, level, unaccelerated flight, it is true that the opposing lift/weight forces are equal. They are also greater than the opposing forces of thrust/drag that are equal only to each other. Therefore, in steady flight: The sum of all upward forces (not just lift) equals the sum of all downward forces (not just weight) ; The sum of all forward forces (not just thrust) equals the sum of all backward forces (not just drag) .

2. Choose your answers from the box and write the letters A) to J) in the blanks.

Listening Script

One of the most significant components of aircraft design is CG. It is the specific point where the mass or weight of an aircraft may be said to center, that is, a point around which, if the aircraft could be suspended or balanced, the aircraft would remain relatively level. The position of the CG of an aircraft determines the stability of the aircraft in flight. As the CG moves rearward, the aircraft becomes more and more dynamically unstable. In aircraft with fuel tanks situated in front of the CG, it is important that the CG is set with the fuel tank empty. Otherwise, as the fuel is used, the aircraft becomes unstable. The CG is computed during initial design and construction, and is further affected by the installation of onboard equipment, aircraft loading, and other factors.

Listening 4   Stalls, Stall Recovery and Spins

1. Answer the questions as below.

Listening Script

After the Lion Air 610 disaster last year, we learned the Boeing 737 Max 8 was equipped with a new system designed to automatically prevent stalls under certain conditions. Pilots had never been trained on or told the system existed. After Ethiopian Airlines Flight 302 crashed, these points have all been revisited. Did the pilots respond properly? Were they aware that deactivating the MCAS (Maneuvering Characteristics Augmentation System) would allow them to regain control of the aircraft? Were they familiar with the process for doing so?

2. Fill in the blanks.

Listening Script

Today I will talk about stalls, stall recovery and spins. An aircraft stall results from a rapid decrease in lift caused by the separation of airflow from the wing's surface brought on by exceeding the critical AOA. A stall can occur at any pitch attitude or airspeed. Stalls are one of the most misunderstood areas of aerodynamics because pilots often believe that an airfoil stops producing lift when it stalls. In a stall, the wing does not totally stop producing lift. Rather, it can not generate adequate lift to sustain level flight. In most straight-wing aircraft, the wing is designed to stall the wing root first. The wing root reaches its critical AOA first making the stall progress outward toward the wingtip. By having the wing root stall first, aileron effectiveness is maintained at the wingtips, maintaining controllability of the aircraft. The wing never completely stops producing lift in a stalled condition. If it did, the aircraft would fall to the earth. The stalling speed of a particular aircraft is not a fixed value for all flight situations, but a given aircraft always stalls at the same AOA regardless of airspeed, weight, load factor, or density altitude.

3. Write T if the statement agrees with the information, and F if the statement contradicts the information.

Listening Script

Your primary consideration after a stall occurs should be to regain positive control of the aircraft. If you do not recover promptly by reducing the angle of attack, a secondary stall and/or a spin may result. A spin may be defined as an aggravated stall which results in the airplane descending in a helical, or corkscrew, path. During aircraft certification tests, normal category airplanes must demonstrate recovery from a one-turn spin or a three-second spin, whichever takes longer.

A secondary stall is normally caused by poor stall recovery technique, such as attempting flight prior to attaining sufficient flying speed.

4. Answer the questions as belew.

Listening Script

If following basic guidelines should be used you encounter a secondary stall, you should apply normal stall recovery procedures. The following basic guidelines should be used to affect a proper stall recovery:

(1) Decrease the angle of attack. Depending on of the type of aircraft, you may find that a different amount of forward pressure on the control wheel is required. Too little forward movement may not be enough to regain lift; too much may impose a negative load on the wing, hindering recovery.

(2) Smoothly apply maximum allowable power. If you are not already at maximum allowable power, increase the throttle to minimize altitude loss and increase airspeed.

(3) Adjust the power as required. As the airplane recovers, you should maintain coordinated flight while adjusting the power to a normal level.

Lesson 5 Aircraft Structure

Listening 1 Cessna 172

1. Fill in the blanks.

Listening Script

The Cessna 172 Skyhawk is an American four-seat, single-engine, high wing, fixed-wing aircraft made by the Cessna Aircraft Company. First flown in 1955, more 172s have been built than any other aircraft. It was developed from the 1948 Cessna 170, using tricycle undercarriage, rather than a tail-dragger configuration.

3. Answer the questions as below.

Listening Script

The structure of the aircraft is called the airframe. Although airplanes are designed for a variety of purposes, most of them have the same major components. The overall characteristics are largely determined by the original design objectives. Most airplane structures include a fuselage, wings, an empennage, landing gear, and a powerplant.

The purpose of the structure is to transfer the lift from the wings to support the weight of the structure and load, and to protect its payload (passengers and cargo). It also has to carry the thrust from the propeller to pull the aircraft through the air. The tail surfaces are attached to provide stability. The control surfaces allow the pilot to override the stability to change the flight path of the aircraft. On the ground the airframe is supported by the undercarriage.

4. Write T if the statement agrees with the information, and F if the statement contradicts the information.

Listening Script

The fuselage is the central body of an airplane and is designed to accommodate the crew, passengers, and cargo. It also provides the structural connection for the wings and tail assembly.

The wings are airfoils attached to each side of the fuselage and are the main lifting surfaces that support the airplane in flight. There are numerous wing designs, sizes, and shapes used by the various manufacturers. Wings may be attached at the top, middle, or lower portion of the fuselage. These designs are referred to as high-, mid-, and low-wing, respectively. The number of wings can also vary. Airplanes with a single set of wings are referred to as monoplanes, while those with two sets are called biplanes.

The landing gear is the principal support of the airplane when parked, taxiing, taking off, or landing. The most common type of landing gear consists of wheels, but airplanes can also be equipped with floats for water operations, or skis for landing on snow. The landing gear absorbs landing loads and supports the airplane on the ground. Typically, the landing gear consists of three wheels—two main wheels which are located on either side of the fuselage and a third wheel positioned either at the front or rear of the airplane. Landing gear employing a rear-mounted wheel is called conventional landing gear. Airplanes equipped with conventional landing gear are sometimes referred to as tail wheel. When the third wheel is located on the nose, it is called a nosewheel. And the design is referred to as tricycle gear.

The powerplant usually includes both the engine and the propeller. The primary function of the engine is to provide the power to turn the propeller.

Listening 2 Empennage Components

2. Complete the table as below.

Listening Script

The empennage includes the entire tail group and consists of fixed surfaces such as the vertical stabilizer and the horizontal stabilizer. The movable surfaces include the rudder, the elevator, and one or more trim tabs. The rudder is attached to the back of the vertical stabilizer. During flight, it is used to move the airplane's nose left and right. The elevator, which is attached to the back of the horizontal stabilizer, is used to move the nose of the airplane up and down during flight. Trim tabs are small, movable portions of the trailing edge of the control surface. These movable trim tabs, which are controlled from the flight deck, reduce control pressures. Trim tabs may be installed on the ailerons, the rudder, and/or the elevator.

Listening 3 The Powerplant

1. Choose the correct letter, A), B) or C).

Listening Script

In the small airplane, the powerplant includes both the engine and the propeller. The primary function of the engine is to provide the power to turn the propeller. It also generates electrical power, provides a vacuum source for some flight instruments, and in most single-engine airplanes, provides a source of heat for the pilot and passengers. The engine is covered by a cowling, or a nacelle, which are both types of covered housings. The purpose of the cowling or nacelle is to streamline the flow of air around the engine and to help cool the engine by ducting air around the cylinders.

The propeller, mounted on the front of the engine, translates the rotating force of the engine into thrust, a forward acting force that helps move the airplane through the air. A propeller is a rotating airfoil that produces thrust through aerodynamic action. A low pressure area is formed at the back of the propeller's airfoil, and high pressure is produced at the face of the propeller, similar to the way lift is generated by an airfoil used as a lifting surface or wing. This pressure differential pulls air through the propeller, which in turn pulls the airplane forward.

Listening 4　Subcomponents

1. Listen to a lecture about subcomponents of an airplane, and fill out the outline as below.

Listening Script

The subcomponents of an airplane include the airframe, electrical system, flight controls, and brakes. The primary function of an aircraft electrical system is to generate, regulate, and distribute electrical power throughout the aircraft. The aircraft's electrical power system is used to operate the flight instruments, essential systems such as anti-icing, etc., and passenger services, such as cabin lighting. The flight controls are the devices and systems which govern the attitude of an aircraft and, as a result, the flight path followed by the aircraft. In the case of many conventional airplanes, the primary flight controls elevators for pitch, ailerons for roll, and the rudder for yaw. These surfaces are operated by the pilot in the flight deck or by an automatic pilot. Airplane brakes consist of multiple pads which place pressure on the rotor which is turning with the wheels. Because airplane brakes are used principally during landings and must absorb enormous amounts of energy, their life is measured in landings rather than miles.

Lesson 6　Flight Instruments Displays

Listening1　Analog Display: the Six Primary Flight Instruments

1. Summary.

Listening Script

If you have an analog watch, it tells the time with hands that sweep around a dial: the position of the hands is a *measurement* of the time. How much the hands move is directly related to what time it is. So if the hour hand sweeps across two segments of the dial, it's showing that twice as much time has elapsed compared to if it had moved only one segment. That sounds incredibly obvious, but it's much more subtle than it first seems. The point is that the hand's movements over the dial are a way of *representing* passing time. It's not the same thing as time itself: it's a representation or an analogy of time. The same is true when you measure something with a ruler. If you measure the length of your finger and mark it on the surface of a wooden ruler, that little strip of wood or plastic you're looking at (a small segment of the ruler) is the same length as your finger. It isn't your finger, of course——it's a representation of your finger: another analogy. That's really what the term analog means.

2. Completion.

Listening Script

There are six primary instruments that have become standard in any instrument panel. Since the early 1970s, these have been arranged in a standard layout referred to as 'the six pack'. They are laid out in two rows of three instruments each. The top row, from left to right, contains the airspeed indicator (ASI), the attitude indicator (AI) and the altimeter (ALT). The bottom row contains the turn coordinator (TC) the directional gyro (DG) and the vertical speed indicator (VSI). A summary of these instruments follows.

3. Fill in the blanks.

Listening Script

The first impression most people have of the cockpit or flight deck is of the complexity and number of instruments. Of all the instruments located in the airplane cockpit, the indicators which provide you information regarding the airplane's attitude, direction, altitude, and speed are collectively referred to as the flight instruments. Traditionally, the flight instruments are sub-divided into categories according to their method of operation. The instruments that reflect your speed, rate of climb or descent, and altitude operate on air pressure differentials and are called pitot-static instruments. A pictorial view of the airplanes' attitude and rate of turn is provided by the attitude indicator and turn coordinator, which operate on gyroscopic principles. The airplane's heading indicator, which also operates using a gyroscope, is usually set by using information from another flight instrument, the magnetic compass.

4. Fill in the table.

Listening Script

The Airspeed Indicator (ASI)

The airspeed indicator (labeled 1) shows the speed at which the aircraft istraveling through the air. In its simplest form, it is nothing more than a spring which opposes the force of the air blowing in the front of a tube attached under the wing or to the nose of the aircraft. The faster the airplane is moving the stronger the air pressure is that acts to oppose the spring and the larger the deflection of the needle from which the pilot reads the craft's speed. Obviously, it's quite a bit more complicated than this, as the pressure exerted by the stream of air varies with the local air density (which continually changes as the airplane climbs or descends), and the ASI must account for this.

The Attitude Indicator (AI)

The attitude indicator (labeled 2) informs the pilot of his or her position in space relative to the horizon. This is accomplished by fixing the case of the instrument to the aircraft and measuring the displacement of the case with reference to a fixed gyroscope inside.

The Altimeter (ALT)

The altimeter (labeled 3) looks somewhat like the face of a clock and serves to display altitude. This is measured by the expansion or contraction of a fixed amount of air acting on a set of springs. As the airplane climbs or descends, the relative air pressure outside the aircraft changes and the altimeter reports the difference between the outside air pressure and a reference, contained in a set of airtight bellows.

Listening 2   Digital Display: the Primary Flight Display (PFD)

2. Answer the questions as below.

Listening Script

On the flight deck, the display units are the most obvious parts of an EFIS system, and are the features which give rise to the name 'glass cockpit' . Now let's discuss about EFIS firstly. An Electronic Flight Instrument System (EFIS) is a flight deck instrument display system in which the display technology used is electronic rather than electromechanical. EFIS normally consists of a primary flight display (PFD) , multi-function display (MFD) and Engine Indicating and Crew Alerting System (EICAS) display. Next, let's take a look at the Primary Flight Display.

The Primary Flight Display (PFD) displays all information critical to flight, including airspeed, altitude, heading, attitude, vertical speed and yaw. The PFD is designed to improve a pilot's situational awareness by integrating this information into a single display instead of six different analog instruments, reducing the amount of time necessary to monitor the instruments. PFDs also increase situational awareness by alerting the aircrew to unusual or potentially hazardous conditions——for example, low airspeed, and high rate of descent——by changing the color or shape of the display or by providing audio alerts.

3. Complete the following table.

Listening Script

The primary flight displays present a dynamic color display of all the parameters necessary for flight path control. The information about airspeed, altitude, vertical speed and soon is provided through the primary flight displays. Airspeed is displayed on a tape and in a digital window on the left side of the primary flight displays. The current Mach number is digitally displayed below the airspeed tape; an airspeed trend vector indicates predicated airspeed in 10 seconds. Selected air speed is displayed above the airspeed tape. Takeoff and landing reference speeds and flap maneuvering speeds are shown along the right edge of the airspeed tape. Maximum and minimum airspeeds are also displayed along the right edge of the airspeed tape.

Altitude is displayed on the altitude tape along the right side of PFD. It is also shown digitally in a window in the middle of the tape. Vertical speed is displayed to the right of the altitude tape with a tape and a pointer.

Listening 3   Magnetic Compass

1. Complete the summary as below.

2. Decide if the following statements (1) ~ (7) are true (T) or false (F).

Listening Script

The magnetic compass was one of the first instruments to be installed in an airplane, and it is still the only direction seeking instrument in many airplanes. If you understand its limitations, the magnetic compass is a reliable source of heading information. The magnetic compass is a self-contained unit does not require electrical or suction power. To determine direction, the compass uses a simple bar magnet with two poles. The bar magnet in the compass is mounted so it can pivot freely and align itself automatically with the earth's magnetic field. The angular difference between the true and magnetic poles at a given point is referred to as variation. Since most aviation charts are oriented to true north and the aircraft compass is oriented to magnetic north, you must convert a true direction to a magnetic direction by correcting for the variation. Deviation refers to a compass error which occurs due to disturbances from magnetic fields produced by metals and electrical accessories within the airplane itself. The remaining error is recorded on a chart, called a compass correction card, which is mounted near the compass.

Chapter 3   Pilot Interview and Flight Training

Lesson 7   Pilot Interview

Listening 1   Flight School Interview

1. Write down the questions.

Listening Script

(1) Tell me about yourself.

(2) Where do you see yourself in five years?

(3) What is your greatest weakness?

(4) What is your greatest strength?

(5) What qualities make a good pilot?

(6) What was the most difficult decision you've made?

(7) What is a pilot's role during an emergency?

(8) Why do you want to be a pilot?

(9) What have you learnt on campus?

(10) How much do you know about our flight academy?

3. Fill in the blanks.

Listening Script

How to make a good performance in your flight school interview? How to prove that you are the certain one? Here are some tips for you.

What is your greatest weakness?

Address this question by minimizing your weaknesses and emphasizing your strengths. Focus on professional traits rather than personal qualities and bring up any education or other opportunities you've used to improve upon your weaknesses. Do not show attempt to show yourself as perfect either. Interviewers want to see if you are self-aware and that you have the ability to improve yourself. Don't highlight a weakness if it is a core competency for the position.

What is your greatest strength?

The interviewer is looking for a strength related to the position. Be prepared to provide examples of your strength and how you've used it in past study. Some examples of strengths to highlight include attention to detail, taking the initiative, positive attitude, working well under pressure, having a desire to learn, problem solving, or working well within a team.

Listening 2   Airline Company Interview

1. Write down the questions.

Listening Script

(1) What is your greatest accomplishment?

(2) What have you done recently to improve as a pilot?

(3) What would your leadership style be as a captain?

(4) Have you ever failed a check-ride or required additional training?

(5) What would you do if a crew member had been drinking?

(6) What is the definition of crew resource management (CRM) ?

(7) What is the importance of checklists and SOP?

(8) What makes you qualified for this position?

(9) How are you different from the other candidates?

(10) Why do you want to work for us?

3. Fill in the blanks.

Listening Script

Employers often seek these qualities in pilots:

Calmness: being able to cope in high pressure situations.

Knowledge of responsibility: pilots know their role in the safety and well being of the employer's customers.

Self motivation: pilots operate separate of supervisors and need to not only motivate themselves but other members of the crew.

Teamwork: you must be able to demonstrate awareness of the role you play within a team outside of piloting duties. A key component of this is effective communication.

Customer service: a good understanding of customer care is important to an employer.

Show an interviewer that you not only meet their current needs but will also be valuable for the company's future or dreams.

Listening 3   Job Interview Tips

1. Completion.

Listening Script

Review the typical job interview questions employers ask and practice your answers. Strong answers are those that are specific but concise, drawing on concrete examples that highlight your skills and back up your resume. While it's important to familiarize yourself with the best answers, it's equally important to listen carefully during your interview in order to ensure your responses give the interviewer the information they are looking for. Also, have a list of your own questions to ask the employer ready. In almost every interview, you'll be asked if you have any questions for the interviewer.

2. Research the Company, and Show What You Know.

Listening Script

Do your homework and research the employer and the industry, so you are ready for the interview question, 'What do you know about this company?' If this question is not asked, you should try to demonstrate what you know about the company on your own. You can do this by tying what you've learned about the company into your responses.

3. Try to Stay Calm.

Listening Script

During the job interview, try to relax and stay as calm as possible. Remember that your body language says as much about you as your answers to the questions. As you answer questions, maintain eye contact with the interviewer. Be sure to pay attention to the question so that you don't forget it, and listen to the entire question before you answer, so you know exactly what the interviewer is asking. If you need to take a moment to think about your answer, that's totally fine, and is a better option than starting out with multiple 'ums' or 'uhs'.

Lesson 8 Flight School

Listening 1 Pilot Study Tips

1. Fill in the blanks.

Listening Script

Decide what you want to fly. FAA's rules for getting a pilot's license (certificate) differ depending on the type of aircraft you fly. You can choose among airplanes, gyroplanes, helicopters, gliders, balloons, or airships. If you are interested in flying ultralight vehicles, you don't need a pilot's license.

You should also think about what type of flying you want to do. There are several different types of pilot's licenses, from student pilot all the way up to airline transport pilot. The information below describes the eligibility, training, experience, and testing requirements for Student Pilots, Recreational Pilots and Private Pilots.

2. Flight Training.

Listening Script

You should recognize the advantages of planning a definite study program and following it as closely as possible. Haphazard or disorganized study habits usually result in an unsatisfactory score on the knowledge test.

The ideal study program would be to enroll in a formal ground school course. This offers the advantages of a professional instructor as well as facilities and training aids designed for pilot instruction. Many of these schools use audiovisual aids or programmed instruction materials to supplement classroom instruction.

If you are unable to attend a ground school, the self-study method can be satisfactory, provided you obtain the proper study materials and devote a reasonable amount of time to study. You should establish realistic periodic goals and, equally important, a target date for completion. Self-discipline is important because it is too easy to 'put off' the study period for some other activity.

3. Complete the chart.

Listening Script

Choosing A Training Provider

You must make your own decision on where to obtain flight training. Once you have decided on a general location, you might want to make a checklist of things to look for in a training provider. By talking to pilots and reading articles in flight magazines, you can make your checklist and evaluate different options. Your choice of a provider might depend on whether you are planning on obtaining a recreational or private certificate or whether you intend to pursue a career as a professional pilot. Another consideration is whether you will train part-time or full-time.

Do not make the mistake of making your determination based on financial concerns alone. The quality of training you receive is very important. Prior to making a final decision, visit the provider you are considering and talk with management, instructors, and students. Evaluate the items on the checklist you developed and then take some time to think things over before making your decision.

Listening 2   Private Pilot

1. Write T if the statement is true, and F if the statement is false.

Listening Script

A private pilot certificate is for those who have always wanted to fly and is the starting point for those who have higher aviation goals. Many private pilots are interested in flying for fun and flying a variety of different aircraft. Others learn to fly so they are able to use an aircraft to support their business or professional interests, like flying themselves to meetings to cut down on travel time. Whatever the goal, private pilots have many options.

A private pilot certificate is the certificate held by the majority of active pilots in the USA. It allows command of any aircraft weighing less than 12, 500 lbs, (subject to appropriate ratings) for any non-commercial purpose and gives almost unlimited authority to fly under visual flight rules (VFR). A private pilot may carry passengers and flight in furtherance of a business. However, a private pilot may not be compensated for services as a pilot. Passengers are only allowed to pay a pro-rata share of flight expenses, such as fuel or rental costs.

2. Fill in the blanks.

Listening Script

With a private pilot certificate, a pilot can rent aircraft from American Flight Schools and fly anywhere in the US. Our diverse fleet enables renters to experience the wide variety of aircraft available in the US market. A private pilot certificate is the minimum requirement for adding other ratings and more advanced certificates. American Flight Schools offers most of the advanced certificates and ratings. Popular advancements include: technically advanced avionics aircraft, high performance aircraft, tailwheel endorsement, mountain flying checkout, and Cirrus training.

3. Answer the questions as below.

Listening Script

What's involved in Flight Training?

Private pilot training involves two main components——ground school and flight training. Ground school is accomplished either in our classroom or through a home study course. It teaches you all the background information that all pilots need to know, but are not easily taught in the air. Topics covered during ground school include: regulations, airplane systems, navigation, aerodynamics, weather theory, and radio communications.

Typically 3-hour lessons are scheduled during which students will log 1-2 hours of flight time. Each lesson is structured to cover the specific maneuvers needed to master in order to earn a certificate. This training prepares the student to take both the written and practical tests. Certification is achieved upon successful completion of these two tests.

What aircraft will you fly?

American Flight Schools has a large diverse Fleet of aircraft in which students can learn to fly. The most common trainers are Cessna 172s, Piper Warriors and Archers, Diamond DA40s and DA42s, as well as a range of other aircraft for aerobatics, glider pilot training, or just because you can. Our fleet of aircraft have options for every person, purpose, and budget.

How long does it take?

That's very much dependent on each student. Flying regularly and studying at home shortens the amount of time in the air and time with an instructor on the ground substantially. Flying a minimum of two, 3-hour lessons per week, a student can expect to have their certificate in approximately 6 months. For people doing flight training for a professional pilot career, contact us for a more detailed training curriculum.

Listening 3　Pilot Training Providers

1. Complete the table.

Listening Script

Pilot training is available on-site at most airports, either through an FAA-certificated (approved) pilot school or through other training providers. An approved school may be able to provide a greater variety of training aids, dedicated facilities, and more flexibility in scheduling. A number of colleges and universities also provide pilot training as a part of their curricula.

＊FAA-approved pilot schools are certificated in accordance with Title 14, Code of Federal Regulations part 141.

Enrollment in an FAA-approved pilot school usually ensures a high quality of training. Approved schools must meet prescribed standards with respect to equipment, facilities, personnel, and curricula. However, individual flight instructors and training companies that are not certificated by the FAA as 'pilot schools' may also offer high quality training, but find it impractical to qualify for FAA certification.

Another difference between training provided by FAA-approved pilot schools and other providers is that fewer flight hours are required to be eligible for a pilot certificate when the training is received through an approved school. The flight hour requirement for a private pilot certificate is normally 40 hours, but may be reduced to 35 hours when training with an approved school. However, since most people require 60 to 75 hours of training, this difference may be insignificant.

2. Matching and completion.

Listening Script

ATP owns and operates the largest flight training fleet in the world, with 406 aircraft. This includes 150 single-engine Piper Archers and 163 Cessna Skyhawk 172s and the largest multi-engine fleet, with 92 Piper Seminoles. All of the Piper Seminoles feature a standardized avionics suite that eases the transition between aircraft. ATP is currently taking delivery of factory-new Piper Archers. Used in the Airline Career Pilot Program, all of the new aircraft are equipped with Garmin G500 glass cockpit avionics suites. All of ATP's Cessna 172s are equipped with a Garmin Moving-Map GPS system. These aircraft are flown for Private Pilot training and the single-engine add-on portion of your training. ATP's Skyhawk fleet continues to grow, accepting delivery of 100 factory-new Cessna 172s through 2023.

Lesson 9   Pilot Certificates

Listening 1   Certificates types

1. Answer the questions.

(1) List the pilot certificates.

(2) Fill in the blanks.

Listening Script

The type of intended flying will influence what type of pilot's certificate is required. Eligibility, training, experience, and testing requirements differ depending on the type of certificates sought.

A private pilot is one who flies for pleasure or personal business without accepting compensation for flying except in some very limited, specific circumstances. The Private Pilot Certificate is the certificate held by the majority of active pilots. It allows command of any aircraft (subject to appropriate ratings) for any noncommercial purpose, and gives almost unlimited authority to fly under VFR. Passengers may be carried, and flight in furtherance of a business is permitted; However, a private pilot may not be compensated in any way for services as a pilot, although passengers can pay a pro rata share of flight expenses, such as fuel or rental costs. If training under 14 CFR part 61, experience requirements include at least 40 hours of piloting time, including 20 hours of flight with an instructor and 10 hours of solo flight.

A commercial pilot may be compensated for flying. Training for the certificate focuses on a better understanding of aircraft systems and a higher standard of airmanship. The Commercial Certificate itself does not allow a pilot to fly in instrument meteorological conditions (IMC), and commercial pilots without an instrument rating are restricted to daytime flight within 50 nautical miles (NM) when flying for hire.

A commercial airplane pilot must be able to operate a complex airplane, as a specific number of hours of complex (or turbine-powered) aircraft time are among the prerequisites, and at least a portion of the practical examination is performed in a complex aircraft. A complex aircraft must have retractable landing gear, movable flaps, and a controllable pitch propeller. See 14 CFR part 61, section 61.31 (c) for additional information.

The airline transport pilot (ATP) is tested to the highest level of piloting ability. The ATP Certificate is a prerequisite for acting as a pilot in command (PIC) of scheduled airline operations. The minimum pilot experience is 1,500 hours of flight time. In addition, the pilot must be at least 23 years of age, be able to read, write, speak, and understand the English language, and be 'of good moral standing'.

2. Get additional knowledge about pilot certificates and fill in the table.

Listening Script

In the US, here are the six types of U.S. pilot certificates a person can obtain. Now I will introduce all of them to you.

Sport Pilot

A sport pilot is the easiest and least restrictive certificate a student can obtain. It's meant for those pilots who wish to fly in light aircraft only, at low altitudes in their local area. Sport pilot certificates can be earned in one of multiple categories: Airplane, powered parachute, glider, rotorcraft and lighter-than-air.

Sport pilots are limited to just one passenger, and are prohibited from flying at night, above 10,000 feet, or in Class B, C or D airspace.

The advantage to a sport pilot certificate is that it only requires 20 hours of logged training time, and most applicants are not required to obtain an FAA medical certificate.

Recreational Pilot

It can be a good option for pilots who want to fly heavier aircraft than those used for sport pilot training, but who don't necessarily want to move on to more advanced training.

The recreational pilot certificate requires at least 30 hours of logged flight time, including 15 hours of dual instruction. Recreational pilots are limited to flights less than 50 nautical miles from their departure airport, can only fly during the day and must stay out of controlled airports (class B, C and D airspace).

Private Pilot

The most common pilot certificate is the private pilot certificate, which is allowed to do much more, like fly at night and at controlled airports.

Private pilots can also fly any aircraft in the category allowed. For instance, most private pilots are certified for the category 'Aircraft—Single Engine Land', hich means they can fly any single engine land airplane (as long as they also have the appropriate ratings and endorsements, like those needed for a high—performance aircraft or complex aircraft.)

Private pilots, like recreational and sport pilots, are not allowed to fly for commercial purposes, and must not be compensated for pilot services.

Private pilot training consists of multiple maneuvers and at least 40 hours of flight time, 20 of which must be with an instructor.

Commercial Pilot

The commercial pilot certificate allows a pilot to be paid for their flying services. Since there are separate regulations for scheduled flights, commercial pilots must also abide by any additional federal aviation regulations pertaining to commercial flying operations.

Becoming a commercial pilot means learning how to fly complex aircraft, which have retractable landing gear, flaps, and a controllable—pitch propeller. Commercial flight training also demands more precision and knowledge about professional flight operations.

Flight Instructor

Becoming a flight instructor involves learning a bit about instructional design, learning theory, and going into all commercial pilot topics much more in depth. The flight instructor certificate allows pilots to share their knowledge of flight with others while gaining necessary experience to move on to an airline.

Airline Transport Pilot

The airline transport pilot (ATP) is the most advanced pilot certificate one can obtain, and it's necessary for those who want to fly commercial airliners for a living. To become eligible for an ATP certificate, a pilot must have logged at least 1,500 hours, be at least 23 years old.

These are the different types of pilot certificates (or licenses, as many people call them.) These should not be confused with ratings or endorsements, which are separate training requirements that allow a pilot with a certain type of certificate to perform additional types of flying, such as instrument flying. You can think of these as extras that are added on to a certificate.

Listening 2　Commercial Pilot Certificate

1. Fill in the blanks.

Listening Script

If you want to be paid to fly in any capacity, a commercial pilot certificate is required.

Training for the certificate takes a pilot back to the basics of airmanship, and away from the rigors of instrument training.

A pilot must obtain ratings for each category and class of airplane she wishes to be paid to operate. For example, if a pilot aspires to fly skydivers in a Cessna 182 and cargo in a King Air C90, she must have commercial certificate with a single engine and amulti-engine rating.

To obtain a commercial certificate in an airplane under FAR Part 61 rules a pilot must have:

(1) 250 hours of flight time, 100 hours of which must be in powered aircraft, and 50 must be in airplanes.

(2) 100 hours of pilot-in-command time, 50 of which must be in airplanes.

(3) 50 hours of cross-country time, 10 of which must be in an airplane.

(4) 20 hours of training, including 10 of instrument, 10 of complex or TAA, and a smattering of cross-country and practical test preparation.

(5) 10 hours of solo training, including a smattering of cross-country and night.

Additional class ratings, such as adding a multiengine rating to single-engine commercial pilot certificate or adding a single-engine rating to a multiengine commercial pilot certificate, will take additional training in that class. FAR 61. 129 has all the specifics.

For pilots who train in an approved FAR Part 141 program, the commercial certificate can be earned with less experience, at a minimum of 190 hours.

Chapter 4   Aircraft System

Lesson 10   Warning and Fire Protection Systems

Listening 1   Visual, Aural and Tactile Warnings

1. Complete the table.

Listening Script

Large transport aircraft have many on-board warning systems used to alert the flight crew to unsafe operational conditions. Some of these systems include fire and overheat, unsafe takeoff, unsafe landing, overspeed, cabin pressure, altitude alert, and ground proximity warning systems. Other warning systems used to provide information to the pilots or the flight engineer can be doors, flight controls, slat position, auto-pilot disconnect and others.

There are many annunciator lights in the aircraft cockpit. In this passage, we'll be looking at lights which warn of malfunction and those lights which are advisory and associated with normal operation. Conditions which require immediate corrective action by the flight crew are indicated by red warning lights located in the area of the pilot's primary field of vision. These lights indicate engine, wheel or APU fires, autopilot disconnect, and landing gear unsafe conditions. Conditions which require timely corrective action by the flight crew are indicated by amber caution lights.

Blue lights inform by crew of electrical power availability, valve position, equipment status and flight attendant or ground crew communication. Blue lights do not require immediate crew attention. When the landing gear is down and locked, the green light illuminated.

2. Complete the outline.

Listening Script

Warnings alert the flight crew to conditions that require action or caution in the operation of the airplane. The character of the signal use varies, depending upon the degree of urgency or hazards involved. Aural, visual and tactile signals are used singularly or in combinations to provide simultaneously, both warning and information regarding the nature of the condition.

Various aural signals call attention to warnings and cautions. An aural warning for airspeed limits is given by aclacker, the autopilot disconnect by a warning tone, cabin altitude by intermittent horn, landing gear positions by a steady horn. The take-off configuration is given by an intermittent horn, and the fire warning by a fire warning bell. Ground proximity warnings and alerts are indicated by voice warnings.

As with the landing and take-off warning system, there are certain configurations in flight which activate the aircraft's stall warning system. The 737 has two independent stall warning systems. Warning an impending stall is required to occur a minimum of 7% above actual stall. Natural stall warning (buffet) usually occurs at a speed prior to stall. In some configurations the margin between stall and stall warning (buffet) is less than the required 7%. Therefore, an artificial stall warning device, a stick shaker, is utilized to provide the required warning.

3. Write T if the statement agrees with the information, and F if the statement contradicts the information.

Listening Script

Flight deck panel annunciator lights are used in conjunction with EICAS messages to help locate and identify affected systems and controls reduce the potential for error. The annunciator lights provide system feedback in response to flight crew action. The lights also assist in fault detection and system preflight configuration when the engines are shut down and to supplement EICAS information. EICAS consolidates engine and subsystem indications and provides a centrally located crew alerting message display. EICAS also displays some system status and maintenance information. They are system alerts, maintenance information; status messages; communication alerts.

Listening 2　Fire Protection System

1. Complete the report.

Listening Script

Washington——The engine of a British Airways jet caught fire in Las Vegas nearly three years ago because a compressor cracked from regular usage, federal investigators announced

Wednesday.

Nobody died in the incident Sept. 8, 2015, but one flight attendant was seriously injured and 19 people suffered minor injuries in the chaotic evacuation, the National Transportation Safety Board found.

After a fire alarm sounded, the captain spent 22 seconds following a checklist before shutting off the fuel supply to the left engine, which allowed an estimated 97 gallons of fuel to spill onto the runway, according to the board.

'If the left engine had been shut down sooner, there would have been less fuel on the runway to feed the fire', the board said in its 30-page report.

2. Complete the table.

Listening Script

If engine fire occurs, you should cancel the warning. Both the fire warning bell cut out switch and the master fire warning light switch can be used to silence the fire alarm bell. Next, you should disengage the auto throttle if engaged and move the respective engine thrust lever until closed fully. You must then move the engine start lever to 'cut off'. Next, you must pull the engine fire warning switch either to the left or the right and hold it for one second to discharge a fire extinguisher bottle.

If the fire warning switch does not extinguish after thirty seconds, you should discharge the second extinguisher bottle by rotating the switch in the opposite direction. After each extinguisher bottle is discharged, the respective bottle discharged light will illuminate. The fire warning light switch will only go out when the fire is extinguished, and the engine compartment temperature decreases. If the fire warning light remains illuminated, you should land at the nearest airport.

3. Dictation.

Listening Script

(1) In the event of APU fire, the APU will automatically shut down.

(2) This action will provide backup for the automatic shutdown feature.

(3) You discharge the extinguishing agent into the APU compartment.

(4) Ensure that the switches are in the normal position.

(5) Test the fire protection system as part of each flight check.

Listening 3   South African Airways Flight 295

1. Answer the questions as below.

Listening Script

South African Airways Flight 295 was a commercial flight from Chiang Kai – shek

International Airport, Taipei, Taiwan to Jan Smuts International Airport, Johannesburg, South Africa, with a stopover in Plaisance Airport, Plaine Magnien, Mauritius. On 28th November 1987, the aircraft serving the flight, a Boeing 747 Combi named Helderberg, experienced a catastrophic in-flight fire in the cargo area, broke up in mid-air, and crashed into the Indian Ocean east of Mauritius, killing all 159 people on board.

The Boeing 747-200B Combi model permits the mixing of passengers and cargo on the main deck according to load factors on any given route and Class B cargo compartment regulations. Flight 295 had 140 passengers and six pallets of cargo on the main deck.

The fire started to destroy the aircraft's important electrical systems, resulting in loss of communication and control of the aircraft. At exactly 00:07 UTC (4:07 local time), the aircraft broke apart in mid-air, the tail section breaking off first, due to the fire beginning to burn the structure of the aircraft, and crashed into the Indian Ocean, about 134 nautical miles (154 miles) from the Airport. Other theories given for the ultimate demise of the aircraft were that the flight crew eventually became incapacitated by the smoke and fire or extensive damage to the 747's control systems rendered the plane uncontrollable before it hit the ocean.

After communication with Flight 295 was lost for 36 minutes, at 00:44 (04:44 local time), Air Traffic Control at Mauritius formally declared an emergency.

In 2014, the South African investigative journalist Mark D. Young presented a theory that a short circuit in the onboard electronics may have caused the fire. The so-called wet arc tracking arises from the action of moisture when the insulation of live wires is damaged. A leakage current to another damaged wire with the respective potential difference may form. The resulting flashover may reach temperatures of up to 5,000℃ (9,000℉). This temperature is sufficient to ignite the thermal-acoustic insulation blankets that were still used until the late 1990s. Such a short circuit caused the fire on board Swissair Flight 111, resulting in the crash of the aircraft in 1998.

Lesson 11   Oxygen System

Listening 1   Passenger Oxygen

1. Listen to the tape, and answer the questions.

Listening Script

A jet blue flight, forced to make an emergency landing. The jet blue A-320 in the air just a few minutes when smoke pours out of the right engine. But it's getting into the cabin, too. The engine blew, smoke started to come throughout the cockpit. Passenger said they heard a loud boom, and smoke began filling the cabin. The flight's oxygen masks failed to drop and had to be manually deployed by the stewards on the flight deck.

A spokesman for JetBlue said the masks are not designed to automatically deploy in these circumstances; they deploy if sensors detect decompression in the cabin.

2. Fill in the blanks.

Listening Script

Cabin pressurization is a process in which conditioned air is pumped into the cabin of an aircraft, in order to create a safe and comfortable environment for passengers and crew flying at high altitudes.

For aircraft, this air is usually bled off from the gas turbine engines at the compressor stage. The air is cooled, humidified, and mixed with recirculated air if necessary, before it is distributed to the cabin by one or more environmental control systems. The cabin pressure is regulated by the outflow valve. An unplanned drop in the pressure of a sealed system, such as an aircraft cabin, and typically results from human error, material fatigue, engineering failure, or impact, causing a pressure vessel to vent into its lower-pressure surroundings or fail to pressurize at all. Such decompression may be classed as Explosive (shown in the picture) , Rapid, or Slow.

3. Now you will hear a piece of news. Try to retell the contents by the questions below.

Listening Script

The aircraft, with 346 passengers and 19 crew members on board, had an explosive decompression event over the South China Sea while cruising at 29,000 feet on a scheduled flight from Hong Kong, to Melbourne, Australia. The event happened about 55 minutes into the flight while the aircraft was over the Pacific Ocean, about 200 miles from Manila. The crew descended to about 10,000 feet above sea level and successfully diverted to Manila. None of the passengers or crew was injured. A portion of the fuselage just forward of the wing root was found missing after the aircraft landed.

Damage included a rupture in the lower right side of the fuselage, just in front of the area where the right wing joins the fuselage. The rupture extended for approximately 2 meters along the length of the aircraft and 1.5 meters vertically. One cylinder associated with the passenger emergency oxygen system, had sustained a sudden failure and forceful discharge of its pressurized contents, rupturing the fuselage and propelling the cylinder upward, puncturing the cabin floor and entering the cabin adjacent to the second main cabin door. The cylinder had impacted the door frame, door handle and overhead paneling, before presumably falling to the cabin floor and exiting the aircraft through the ruptured fuselage.

4. Dictation.

Listening Script

(1) The passenger's oxygen can be operated both manually and automatically.

(2) On the left side of the cabin there are four oxygen masks to each generator.

(3) It would be necessary to use the passenger oxygen system in the event of cabin depressurization.

(4) The passenger oxygen switch is located on the aft overhead panel.

(5) Passenger oxygen must be used when smoke is generated.

5. Choose the correct letter, A), B) or C).

Listening Script

(1) Passenger oxygen is chemically generated and is completely separated from the flight crew oxygen.

Question: what kind of oxygen does the passenger oxygen system use?

(2) Once in operation, the oxygen flows for approximately twelve minutes.

Question: how long does the passenger oxygen flow once in operation?

(3) When the oxygen mask doors are electrically opened and masks are developed, the passenger oxygen ON light illuminates.

Question: in what situation does the passenger oxygen ON light illuminate?

(4) A mask drops in front of each passenger if cabin altitude climbs to 14,000 feet.

Question: in what situation does a passenger oxygen mask drop?

(5) It would be necessary to the passenger oxygen system in the event of cabin depressurization.

Question: when would it be necessary to use the passenger system?

Listening 2   Flight Crew Oxygen

1. Write T if the statement agrees with the information, and F if the statement contradicts the information.

Listening Script

The flight crew oxygen system is completely separate from the passenger oxygen system. It uses a quick-donning diluter demand mask / regulators located at each crew station. Oxygen is supplied by a single cylinder. Pressure is read on the indicator located on the aft overhead panel when the Battery Switch is ON. Oxygen flow is controlled through a pressure-reducing regulator to supply low pressure oxygen to a shut-off valve located behind the First Officer's seat. Normal pressure is 1,850 P. S. I. A quick-donning mask is located within easy reach crew members. Oxygen flow is controlled by a diluter-demand type regulator located immediately adjacent to each crew station.

2. Fill in the blanks.

Listening Script

Two independent oxygen systems are provided, one for the flight crew and one for the passengers. Portable oxygen cylinders are located throughout the airplane for emergency use. The flight crew oxygen system uses quick-donning masks and regulators located at each crew station.

Oxygen pressure is displayed on the lower EICAS status display. Flight crew and observer masks and regulators are installed in oxygen mask panels near each seat. Squeezing the red oxygen mask release levers releases the mask from stowage.

3. Complete the picture below.

Listening Script

The flow indicator is located on the top right of the stowage box and shows a yellow cross when oxygen is flowing. When the left stowage box door is shut, the oxygen ON flag remains in view indicating that oxygen is still available to the regulator. With the mask not stowed and stowed box door closed operating the reset/test slide lever shuts off oxygen flow to the regulator and the oxygen ON flag is removed. With the left door closed pushing the reset/test lever will cut off oxygen flow to the regulator. The reset / test slide lever is located on the left side of the box. To start the flow of oxygen again, the left door must be opened.

4. Fill in the blanks.

Listening Script

Pilot requirements: Use of oxygen

A) Unpressurized aircraft.

Each pilot of an unpressurized aircraft shall use oxygen continuously when flying at altitudes above 10,000 feet through 12,000 feet MSL. For that part of the flight at those altitudes that is of more than 30 minutes duration; and above 12,000 feet MSL.

B) Pressurized aircraft.

Whenever a pressurized aircraft is operated with the cabin pressure altitude more than 10,000 feet MSL, each pilot shall comply with paragraph of this section.

Whenever a pressurized aircraft is operated at altitudes above 25,000 feet through 35,000 feet MSL, unless each pilot has an approved quick-donning type oxygen mask——At least one pilot at the controls shall wear, secured and sealed, an oxygen mask that either supplies oxygen at all times or automatically supplies oxygen whenever the cabin pressure altitude exceeds 12,000 feet MSL; and during that flight, each other pilot on flight deck duty shall have an oxygen mask, connected to an oxygen supply, located so as to allow immediate placing of the mask on the pilot's face sealed and secured for use.

(3) Whenever a pressurized aircraft is operated at altitudes above 35,000 feet MSL, at least one pilot at the controls shall wear, secured and sealed, an oxygen mask required.

(4) If one pilot leaves a pilot duty station of an aircraft when operating at altitudes above 25,000 feet MSL, the remaining pilot at the controls shall put on and use an approved oxygen mask until the other pilot returns to the pilot duty station of the aircraft.

Listening 3　Portable Oxygen Cylinders

1. Answer the questions as below.

Listening Script

Helios Airways flight HCY522

On 14 August 2005, a Boeing 737-300 aircraft, operated by Helios Airways, departed Larnaca, Cyprus at 06:07 for Prague, Czech Republic, via Athens, Hellas. The aircraft was cleared to climb to FL340 and to proceed direct to RDS VOR. As the aircraft climbed through 16,000ft, the Captain contacted the company Operations Centre and reported a Take-off Configuration Warning and an Equipment Cooling system problem. Several communications between the Captain and the Operations Centre took place in the next eight minutes concerning the above problems and ended as the aircraft climbed through 28,900ft. Thereafter, there was no response to radio calls to the aircraft. During the climb, at an aircraft altitude of 18,200ft, the passenger oxygen masks deployed in the cabin. The aircraft leveled off at FL340 and continued on its programmed route. During the sixth holding pattern, the Boeing 737 was intercepted by two F-16 aircraft of the Hellenic Air Force. One of the F-16 pilots observed the aircraft at close range and reported at 08:32 that the Captain's seat was vacant, the First Officer's seat was occupied by someone who 2 was slumped over the controls, the passenger oxygen masks were seen dangling and three motionless passengers were seen seated wearing oxygen masks in the cabin. No external damage or fire was noted and the aircraft was not responding to radio calls. At 08:49, he reported a person not wearing an oxygen mask entering the cockpit and occupying the Captain's seat. At 11:49, flight attendant Andreas Prodromou entered the cockpit and sat down in the captain's seat, having remained conscious by using a portable oxygen supply. Prodromou held a UK Commercial Pilot License, but was not qualified to fly the Boeing 737. Crash investigators concluded that Prodromou's experience was insufficient for him to gain control of the aircraft under the circumstances.

Lesson 12　APU System

Listening 1　Definition

1. Fill in the blanks.

Listening Script

The auxiliary power unit (APU) is a self - contained gas turbine engine located in the airplane tail cone. The APU air inlet door is located between the horizontal and vertical stabilizers on the right side of the airplane. While the primary purpose of the APU is to supply electrical power and bleed air on the ground before engine starts, the APU can also be started in flight. In flight, APU bleed air is available up to approximately 17, 000 feet.

2. Label parts of the APU given in the column on the left side.

Listening Script

Now let's look at a picture. This is a picture illustrating the location of different parts of an APU. The number one arrow points to an air duct, the number two arrow refers to an accessory cooling air duct. Ordinarily, from number three to seven, are an air diffuser duct, air inlet door, APU bleed air duct, APU fuel line and exhaust duct.

Listening 2　APU System Description

1. Dictation.

Listening Script

(1) The APU is started either by an electric start motor or a turbine starter.

(2) On the ground, the electric starter is limited to three consecutive starts.

(3) The APU is a self-contained gas turbine engine.

(4) It contains of two-stage compressor, a turbine and an accessory drive section.

(5) Pressure is sensed in the bleed feeder duct.

2. Complete the outline.

Listening Script

The APU is started either by an electric start motor or a turbine starter. The electric starter is powered by the APU battery. The main airplane battery powers the inlet door, fuel valve, and fire detection system. The air turbine starter uses engine bleed air or ground cart air to start the APU. Starter selection is automatic. The air turbine starter has priority over the electric start motor when there is sufficient bleed air duct pressure. APU fuel is supplied from the left fuel manifold by any operating of the AC fuel pump or the DC fuel pump. With AC power available and the APU selector in the ON Position, the left forward fuel pump operates automatically. If AC power is not available or no AC pump pressurizes the left fuel manifold, the DC pump in the left main tank provides APU fuel. When the APU air inlet door reaches the full position, the starter engages. After the APU reaches the proper speed, ignition and fuel are provided. When the APU reaches approximately 50 percent, the starter disengages and ignition is turned off.

Listening 3   APU Operation

1. Complete the table.

Listening Script

Now let's look at the illustration of a forward overhead panel. One the top of it there are four small display lights. The first on the left is blue. The rest are all amber. From the left to the right, the first one is an APU low oil quantity light. It illuminates if APU oil quantity is insufficient for extended operation. The second one is an APU low oil pressure light. It will illuminate when APU oil pressure is low, causing the APU to initiate an automatic shutdown. The light will keep illuminated during start until the APU oil pressure is normal. The third one is an APU high oil temperature light. It will illuminate when the APU oil temperature is excessive, causing the APU to initiate an automatic shutdown. The forth is an APU overspeed light. Under the four small lights there are two dashboards. The one with figures from 0 to 8 is an APU exhaust temperature indicator. The other one is an APU generator AC Ammeter. Under the APU generator AC Ammeter is an APU switch with three positions: OFF, On and Start.

Listening 4   Korean Airlines Flight 926
Listening Script

A Korean Airlines Boeing 747−400, registration HL7488 performing flight KE−907 from Seoul to London Heathrow, was enroute near Irkutsk (Russia) when the crew received an APU fire indication, declared emergency and diverted to Irkutsk for a safe overweight landing. Tower reported seeing no smoke or fire during the roll out, attending emergency services found no trace of fire, heat or smoke. The airplane was examined, the indication was determined false. Following repairs the airplane was released to continue the flight and is estimated to reach London with a delay of 4 hours.

Lesson 13   Fuel System

Listening 1   Avoiding A Common Preflight Mistake

2. Fill in the blanks.

Listening Script

Many years ago, I owned a Cessna 195, and at that time I was based at a little airport in the mountains of California… it was a little short airport with nothing but rocks and trees on both ends of the runways. Certainly there was no chance for a safe landing if you had an engine problem right after takeoff.

I arrived early one day, just to exercise the airplane, and go for a little scenic flight, practice my landings. That aircraft had the three standard fuel drains, two fuel tanks and a fuel sump. During my preflight, I drained the right wing sump. I drained the fuel sump at the engine. But I forgot to drain the left wing tank. Now I'd owned this airplane for several years and I had never gotten one drop of water out of that airplane. I drained it religiously, never had a problem.

I got in the airplane, fastened my seatbelt, then I pulled out my checklist, my pre-flight checklist, that is. A little bit late, but at least I pulled it out. I went through it and I realized I'd forgotten to drain the left tank. By now I'm in the airplane, the doors closed, the seatbelt's on. I've never gotten any water before. Why should I go through the trouble of doing this? But fortunately I decided, hey I'm in no hurry. Why not just follow good discipline and good practice and do it?

I got out of the airplane. I got my fuel drainer. I drained a cupful of fuel. I looked at it and discovered it was all water. I drained another cupful, all water. By now I was getting goose bumps. I drained 17 cups of water out of that tank.

Listening 2   Fuel System in Small Aircraft

1. Fill in the blanks.

Listening Script

The fuel system is designed to provide an uninterrupted flow of clean fuel from the fuel tanks to the engine. The fuel must be available to the engine under all conditions of engine power, altitude, attitude, and during all approved flight maneuvers. Two common classifications apply to fuel systems in small aircraft: gravity-feed and fuel-pump systems.

The gravity-feed system utilizes the force of gravity to transfer the fuel from the tanks to the engine.

For example, on high-wing airplanes, the fuel tanks are installed in the wings. This places the fuel tanks above the carburetor, and the fuel is gravity fed through the system and into the carburetor.

If the design of the aircraft is such that gravity cannot be used to transfer fuel, fuel pumps are installed. For example, on low-wing airplanes, the fuel tanks in the wings are located below the carburetor. Both gravity-feed and fuel-pump systems may incorporate a fuel primer into the system.

2. Answer the questions as below.

Listening Script

The fuel primer is used to draw fuel from the tanks to vaporize fuel directly into the cylinders prior to starting the engine. During cold weather, when engines are difficult to start, the fuel primer helps because there is not enough heat available to vaporize the fuel in the carburetor. It is important to lock the primer in place when it is not in use. If the knob is free to move, it may vibrate out during flight and can cause an excessively rich mixture. After leaving the fuel tank and before it enters the carburetor, the fuel passes through a strainer which is used to removes any moisture and other sediments in the system.

3. Cloze.

Listening Script

Since these contaminants are heavier than aviation fuel, they settle in a sump at the bottom of the strainer assembly. A sump is a low point in a fuel system and/or fuel tank. The fuel system may contain sump, fuel strainer, and fuel tank drains, which may be collocated.

The fuel strainer should be drained before each flight. Fuel samples should be drained and checked visually for water and contaminants. Water in the sump is hazardous. Because in cold weather the water can freeze and block fuel lines. In warm weather, it can flow into the sump and stop the engine. If water is present in the sump, more water in the fuel tanks is probable and they should be drained until there is no evidence of water. Never take off until all water and contaminants have been removed from the engine fuel system.

Aircraft with fuel-pump systems have two fuel pumps. The main pump system is engine driven with an electrically driven auxiliary pump provided for use in engine starting and in the event the engine pump fails. The auxiliary pump, also known as a boost pump, provides added reliability to the fuel system. The electrically driven auxiliary pump is controlled by a switch in the flight deck.

Listening 3   Fuel System Description

1. Complete the summary.

Listening Script

The fuel system supplies fuel to the engine and APU. The fuel is contained in three tanks located with the wings and wing center section. Tank No. 1 and No. 2 are integral with the wing structure. The center tank lies between the wing roots within the fuselage area. Each fuel tank contains two AC powered fuel pumps. A single pump can supply sufficient fuel to operate one engine under all conditions. Two center tank pumps are override jettison pumps. These pumps have a higher output pressure than the left and right main rank fuel pumps. The center tank pumps override the main tank pumps so that center tank fuel is used before wing tank fuel. If a center pump has low output pressure (within more than 40 pounds of fuel remaining), the fuel pump switch press light illuminates and the Fuel Pump Center EICAS advisory message is displayed. If there is less than 400 pounds of fuel remaining, the Fuel Low Center EICAS advisory message is displayed.

2. Dictation.

Listening Script

(1) Engine fuel shutoff valves are located at each engine-mounting wing station.

(2) The engine fuel manifolds are interconnected by use of the cross feed valve.

(3) Check valves are located throughout the fuel system to ensure the proper direction of fuel flow and to prevent transfer of fuel between tanks.

(4) Center tank check valves open at a lower differential pressure than the check valves in the No. 1 and No. 2 main tanks.

(5) This ensures that center tank fuel is used before main tank fuel, even though all fuel pumps are operating.

Listening 4   Defueling

1. Fill in the blanks.

Listening Script

Fuel dumping, or a fuel jettison, is a procedure used by aircraft in certain emergency situations before a return to the airport shortly after takeoff, or before landing short of its intended destination to lighten aircraft's weight.

Aircraft have two major types of weight limits: the maximum takeoff weight and the maximum structural landing weight, with the maximum structural landing weight always being the lower of the two. This allows an aircraft on a normal, routine flight to take off at the higher weight, consume fuel en route, and arrive at a lower weight. If a flight takes off at the maximum structural takeoff weight and then faces a situation where it must return to the departure airport due to certain mechanical problems, or a passenger medical issue, there will not be time to consume the fuel meant for getting to the original destination, and the aircraft may be over the maximum structural landing weight to land back at the departure point.

Listening 5   Avianca Flight 52

1. Answer the questions.

Listening Script

Minimum Fuel Emergency

Bob: John, in low-fuel situations pilots are often reluctant to declare an emergency and they'll use the term, Minimum fuel. Is Minimum fuel going to get you on the ground any faster?

John: No. Minimum fuel will not, and that's very important to understand. It is not a declaration of an emergency. It's merely a statement that is saying that you cannot accept any holding or other delays. It also means to ATC that you do have enough fuel to make it to your destination.

Bob: If you're sitting there sweating bullets on your fuel, you want to be using the real word and declare an emergency.

John: Declare an emergency. Absolutely. Declare an emergency. There have been numerous accidents over the years that have illustrated that point. Probably the most famous is the Avianca at JFK several years ago.

Bob: The pilots inside the cockpit were talking about their situation. The Captain told the First Officer to declare an emergency. The 'E' word never left the cockpit and the airplane crashed short of the airport, killing numerous people.

John: Exactly. Right.

2. Listen to a piece of news and write the numbers 1～10 next to the words A) to J) as below to show the order in which you hear them.

Listening Script

On January 25, 1990, Avianca Flight 52 was much delayed in approaching its destination due to congestion and bad weather. It had been in a holding pattern off the coast near New York for over one hour due to fog and wind interfering with smooth arrivals and departures into John F. Kennedy International Airport. During this hold the aircraft was exhausting its reserve fuel supply, which would have allowed it to divert to its alternate, Boston, in case of an emergency or other critical situation. When first put on hold, the crew of Flight 52 thought that they would be landing soon, after a few aircraft also on hold in front of them had landed. The bad weather, wind shear and other factors caused the pilots of these aircraft to abort their landings, and the hold time increased.

Twenty seven minutes after entering the hold, New York air traffic control (ATC) asked the crew how long they could continue to hold, to which the first officer replied, 'about five minutes'. The first officer then stated that their alternate was Boston, but since they had been holding for so long they would not be able to make it there anymore. Even though Flight 52 had fuel issues, ATC passed the flight to another person, presumably unaware there was any urgency to landing this airplane. The new controller then cleared the aircraft for an approach to runway 22L and informed the flight of wind shear at 1,500 feet. As Flight 52 flew the ILS approach, they encountered wind shear at an altitude of less than 500 feet. As a result, the plane descended below the planned glide slope and almost crashed into the ground short of the runway. The pilots were forced to abandon the landing, even though they knew the plane did not have enough fuel to turn around for another attempt. The crew alerted the controller that they were low on fuel, and in a subsequent transmission stated, 'We're running out of fuel, sir'. The controller then asked the crew to climb, to which the first officer replied, 'No, sir, we're running out of fuel'.

Moments later, with the airplane still very close to the ground, the number four engine flamed out, shortly followed by the other three. With the aircraft's main source of electrical power——generators driven off, it would have caused many nonessential electrical systems to lose power. The cabin thus would have been plunged into darkness. With no engine thrust, the plane lost height.

The aircraft struck the ground and slid down a hill in the town, splitting into two pieces as it reached the bottom. The impact snapped off the cockpit, which landed over 100 feet away in the side of an unoccupied house. Eighty-five people survived the crash with injuries, while 73 passengers and crew died.

Lesson 14   Anti-Icing System

Listening 1   Anti-ice System Description

1. Answer the questions as below.

(1) Snow and ice on these areas in essence changes their shape and disrupts the airflow across the surface, hindering the ability to create lift.

(2) We use deicing fluid, a mixture of a chemical called glycol and water, is generally heated and sprayed under pressure to remove ice and snow on the aircraft.

Listening Script

A plane's wings and rear tail component are engineered with a very specific shape in order to provide proper lift for flight. Snow and ice on these areas in essence changes their shape and disrupts the airflow across the surface, hindering the ability to create lift.

Whenever snow, ice, or even frost has accumulated on the aircraft, the pilots call on the airport deicing facility to have it removed. Deicing fluid, a mixture of a chemical called glycol and water, is generally heated and sprayed under pressure to remove ice and snow on the aircraft.

While it removes ice and snow, deicing fluid has a limited ability to prevent further ice from forming. If winter precipitation is falling, such as snow, freezing rain or sleet, further action needs to be taken to prevent ice from forming again on the aircraft before takeoff.

3. Complete the table.

Listening Script

Thermal anti-icing, electrical anti-icing, and rain repellent are the systems provided for ice and rain protection. The engine bleed air thermal anti-icing prevents the formation of ice on the wing leading edge, engines and engine cowls. The flight deck windows, air data sensors, drain masts, and potable water lines have electric anti-icing systems. The ice detection system automatically operates these systems during icing conditions.

4. Complete the summary.

Listening Script

The wing anti-icing system prevents ice on slats two, three, four, ten, eleven, and twelve. It uses air from the pneumatic system. The system only operates with the airplane in the air. It comes on when the ice detection system finds ice or the flight deck switch is set to the ON position. There is a pressure regulating and shutoff valve and a pressure sensor inside the leading edge of each wing. A perforated spray tube carries the pressure regulated air into the slats. The air heats the slats and then goes overboard through a vent in the slat.

The engine anti-ice system use engine bleed air to prevent ice on the forward edge of the engine cowl. Each engine has three anti-icing valves, the left and right valves and cowl valves. The cowl valve provides thirteenth stage bleed air to the nose cowl lips. These valves open when the respective engine anti-ice switch is placed ON. The Valves Open lights illuminate bright while the respective valves are in transit and then illuminate dim when the respective valves are open. Engine anti-ice operation is controlled from the cockpit by individual engine anti-ice switches. The engine anti-ice system may be operated on the ground and in-flight. Engine anti-ice system must be on during all ground and flight operations when icing conditions exist or are anticipated.

Listening 2   Ice Detection System

1. Choose the correct letter, A), B) or C).

Listening Script

The ice detection system has two ice detectors, one on each side of the forward fuselage. The ice detection system operates the wing and engine anti-ice system automatically when the airplane is in the air. The detector sends a signal to the wing and engine anti-ice systems when it finds ice on the detector. This causes the anti-ice systems to come on. Each air data probe has an electrical heater element. The electrical load management system controls the power to each heater element. On the ground with both engines off, the system does not heat any probes. On the ground with one or both engines running, these conditions occur. Firstly, total air temperature probes are not heated. Secondly, pitot probes are on low heat. Thirdly, angle of attack sensors are on full heat. In flight, the system fully heats all probes and sensors.

2. Write down the summary.

Listening Script

Updrafts in a thunderstorm support abundant liquid water with relatively large droplet sizes. When carried above the freezing level, the water becomes supercooled. When temperature in the upward current cools to about −15℃ , much of the remaining water vapor sublimates as ice crystals. Above this level, at lower temperatures, the amount of supercooled water decreases. Supercooled water freezes on impact with an aircraft. Clear icing can occur at any altitude above the freezing level, but at high levels, icing from smaller droplets may be rime or mixed rime and clear ice. The abundance of large, supercooled water droplets makes clear icing very rapid between 0℃ and −15℃ encounters can be frequent in a cluster of cells. Thunderstorm icing can be extremely hazardous. Thunderstorms are not the only area where pilots could encounter icing conditions. Pilots should be alert for icing anytime the temperature approaches 0℃ and visible moisture is present.

Lesson 15　Hydraulic System

Listening 1　Hydraulic System Description

1. Answer the questions as below

Listening Script

Hydraulics is often used on small airplanes to operate wheel brakes, retractable landing gear, and some constant−speed propellers. On large airplanes, hydraulics is used for flight control surfaces, wing flaps, spoilers, and other systems. For small aircraft, A basic hydraulic system consists of a reservoir, pump (either hand, electric, or engine driven) , a filter to keep the fluid clean, selector valve to control the direction of flow, relief valve to relieve excess pressure, and an actuator.

2. Dictation.

Listening Script

(1) Hydraulic power is one of the power sources which widely used in the modern airplanes to operate various airplane units and mechanisms.

(2) The hydraulic system of the modern airplanes performs many functions.

(3) Among the units commonly operated by hydraulic systems are landing gear, wing flaps, slats, wheel brakes, nose, wheel steering, and primary flight control surfaces.

(4) The engine−driven pump which is directly coupled to the engine will operate when the engine is running.

(5) Hydraulic systems operate independently at 3,000 Psi normal pressures.

3. Complete the table.

Listening Script

There are three hydraulic systems which operate independently at 3,000 Psi normal pressure. The three system are named left, center and right for the location of their maincomponets. Each system has its own reservoir, pump and filter. The left system has an engine−driven pump and an alternating current motor pump. The right AC bus supplies power to alternating current motor pump. The left system supplies power to flight controls and left thrust reverser. The right system also has an engine-driven pump and an alternating current motor pump. The right hydraulic system supplies power to the light controls, the normal main gear brakes, and the right thrust reverser. The center system has two alternating current motor pumps, two air-driven pumps and a ram air turbine pump. The left and right AC buses supply power to both alternating current motor pumps. Pneumatic power from the two engines or the auxiliary power unit power operates the air-driven pumps. The center system supplies power to the flight controls, the leading edge slats, the trailing edge flaps, the alternate reverser, main gear brakes and so on.

Listening 2   Landing Gear System Description

2. Complete the table.

Listening Script

Now let discuss the advantages and disadvantages of tricycle gear airplane and tailwheel gear airplane. A tricycle gear airplane has three advantages:

(1) It allows more forceful application of the brakes during landings at high speeds without causing the aircraft to nose over.

(2) It permits better forward visibility for the pilot during takeoff, landing, and taxiing.

(3) It tends to prevent ground looping by providing more directional stability during ground operation since the aircraft's center of gravity is forward of the main wheels. The forward CG keeps the airplane moving forward in a straight line rather than ground looping.

Tail wheel landing gear aircraft have two main wheels attached to the airframe ahead of its CG that support most of the weight of the structure. A tail wheel at the very back of the fuselage provides a third point of support. This arrangement allows adequate ground clearance for a larger propeller and is more desirable for operations on unimproved fields.

With the CG located behind the main gear, directional control of this type aircraft becomes more difficult while on the ground. This is the main disadvantage of the tail wheel landing gear. For example, if the pilot allows the aircraft to swerve while rolling on the ground at a low speed, he or she may not have sufficient rudder control and the CG will attempt to get ahead of the main gear which may cause the airplane to ground loop. Lack of good forward visibility when the tail wheel is on or near the ground is a second disadvantage of tail wheel landing gear aircraft. These inherent problems mean specific training is required in tail wheel aircraft.

3. Choose the correct letter, A), B) or C).

Listening Script

The landing gear system supports the airplane during takeoff, landing, taxiing, and when parking. These ground operation require that the landing gear be capable of steering, braking, and absorbing shock. The airplane has two main landing gears and a single nose gear. The nose gear is a conventional steerable two wheel unit. Each main gear has six wheels in tandem-par. To improve turning radius, the aft axle of each main gear is steerable. The main gears are located inboard of each engine, aft of the rear wing spar. The nose gear is located below the aft bulkhead of the flight compartment.

Question (1) Which of the following is not mentioned as a capability of ground operations?

Question (2) According to the passage, how many wheels do the landing gears have?

Question (3) Why is the aft axle of each main gear steerable?

Question (4) Where is the nose gear located?

4. Complete the summary.

Listening Script

The landing gears are normally controlled by the landing gear lever. On the ground, the lever is held in ON position by an automatic lever lock. The lever lock can be manually overridden by pushing and holding the landing gear lever lock override switch. In flight, the lever lock is automatically released through air/ground sensing. Hydraulic power for retraction, extension, and steering is supplies by the center hydraulic system. An alternate extension system is also provided. The brake hydraulic system is powered by the right hydraulic system. The alternate brake system is powered by the center hydraulic system.

Antiskid protection is provided with both systems, but the autobrake system is available only through the normal system. A brake temperature monitor system and tire pressure indicator system displays each brake temperature and tire pressure on the gear synoptic display.

Listening 3　Eastern Air Lines Flight 401

1. Complete the table.

Listening Script

A Lockheed L-1011 crashed at 2342 Eastern Standard Time, December 29th, 1972, approximately 18 miles west northwest of Miami International Airport, Miami, Florida. There were 176 people aboard the aircraft including 8 crewmembers. Among them, 99 people were fatally injured. The plane did not land first time because the nose landing gear position indicating system on the aircraft did not indicate that the nose gear was locked in the down position. The possible cause of this accident was the failure of the flight crew to monitor the flight instruments during the final 4 minutes of flight, and to detect an unexpected descent soon enough to prevent impact with the ground.

Lesson 16   Communication System

Listening 1   Radio Communication System

1. Complete the Chart.

Listening Script

The radio communication system consists of the very high frequency radio communication system, the high frequency radio communication system, and selective calling system.

VHF airborne communication sets operate in the frequency range from 108.0 MHz to 135 MHz. Many VHF radios have the transmitter, receiver, power supply, and operating controls built into a single unit. This unit is frequently installed in a cutout in the instrument panel.

There are three independent VHF voice / data radios, which are designated VHF L (left), VHF C (center) and VHF R (right). Any VHF radio can be controlled by any radio tuning panel. The audio control panels are used to control voice transmission and receive monitoring.

VHF L is configured for voice communication only.

VHF C and VHF R can be configured for data or voice communication only. However, only one VHF radio can operate in the data mode at a time. Normally VHF C is configured for data communication.

A high frequency communication system is used for long-range communication. HF systems operate essentially the same as a VHF system, but operate in the frequency range from 3 MHz to 30 MHz. Communications over long distance are possible with HF radio because of the longer transmission range. HF transmitters have higher power outputs than VHF transmitters.

There are two independent HF communication radios, designated HF L and HF R. Each HF radio can be tuned by any radio tuning panel. The selective calling system monitors the three VHF radios and two HF radios. When the system receives a call from the ground station, the crew is alerted through the communication crew alerting system.

Listening 2   Communicating

1. Fill in the blanks.

Listening Script

The connection between your voice and the radio is via a headset which contains the boom microphone for speaking, and earphones for listening. The headset has two plugs which are connected to a socket on the instrument panel or the centre console. All training aeroplanes are required to have headsets and an electronic intercom for communications within the aircraft. There will be a small intercom control panel with two knobs – one for volume and one for squelch. The squelch controls the sensitivity of the microphones. If the squelch is turned down, the background noise is removed.

2. Match the situations (1) ~ (7) with the solutions A) ~ G).

Listening Script

What if the radio doesn't work?

If the radio doesn't work, carry out the following checks:

(1) If there is no noise in the headset, check earphone volumes and plug connections, check intercom on and volume and squelch is up.

(2) If you hear the instructor okay but not yourself, check the microphone plug is fully in.

(3) If you have to puff to trigger voice, the intercom squelch is set too low.

(4) If the intercom works okay but there is no radio reception, select the audio selector panel to phones and correct radio. Check radio is on, avionics master is on, radio on correct frequency, radio squelch is set.

(5) If still no reception, check second radio on same frequency.

(6) If the reception is weak, check squelch and volume and try other radio.

(7) If there is no reply to your transmissions, note if there is a side-tone, which is a change in the nature of your voice and a slight hiss as you talk with the transmit button depressed. If no side-tone and the intercom is normal, try the second radio. If you can only receive the intercom then use the hand-held mike to talk to the tower.

3. Answer the questions.

Listening Script

The microphone is like a telephone with the important following distinctions:

(1) The transmit button must be depressed for you to transmit;

(2) The transmitting, most radio sets are unable to receive, and;

(3) Only one transmission from one station within range can occur on the frequency in use without interference. While you are transmitting, no one else can.

4. Write T if the statement agrees with the information, and F if the statement contradicts the information.

Listening Script

Your radio will continue to transmit as long as the transmit button is depressed. Even if you are not speaking, the carrier wave will continue to be transmitted, blocking out other stations that are trying to call on that particular frequency. So, at the end of your transmission ensure that the PTT is released.

Before transmitting, there are some basic rules which you must always follow:

(1) Listen out on the frequency to be used and avoid interference with other transmissions. If another station , be it another aircraft or an air traffic services (ATS) unit, is transmitting and if a response to its call is awaited do not interrupt. Be particularly aware of this in the rare situation of hearing a distress (Mayday) call, or an urgency (pan-pan) call. Of course, if you wish to transmit a distress or urgency call then you are entitled to interrupt any transmissions of lower priority.

(2) Decide what you want to say. For most communications, we use a standard radio phraseology which is easily learnt. This simplifies things for both the person transmitting and the person receiving. Avoiding long silences and hesitation sounds (ah, um, er, etc.) during your transmission but do use pauses where appropriate. Don't use slang and don't clutter the frequency with social chat.

(3) Have the wording of your intended transmission clearly in your mind (or even written down) prior to pressing the transmit button will help you to avoid hesitation.

5. Complete the table.

Listening Script

When using a microphone, make sure of the following:

(1) Actuate the press-to-talk (PTT) switch before commencing to talk, and do not release it until after your message is completed.

(2) Speak with the microphone close to or just touching your upper lip.

(3) Do not significantly vary the distance between your lips and the microphone.

(4) Speak directly into the microphone.

(5) Speak a little slower than normal, but at normal volume.

(6) Do not raise your voice or shout, and do not speak in a whisper.

(7) Pronounce each word clearly and ensure that you clearly annunciate the end of the word. Running words together, or slurring them, may make reception difficult.

(8) Pause briefly before and after the transmission of numbers.

Listening 3   Interphone Communication System

1. Fill in the blanks.

Listening Script

Given-Boeing 757, the Interphone Communication System includes the:

(1) Flight Interphone System

(2) Cabin Interphone System

(3) Service Interphone System

(4) Passenger Address System

The interphone systems allow the flight crew to communicate with the flight attendants, ground personnel and maintenance technicians. The Passenger Address (PA) system allows flight attendants or the flight crew to make announcements in the passenger cabin.

The Flight interphone, Service interphone, Cabin interphone and PA systems are normally operated through the Audio Control Panel in conjunction with the Pilot Call Panel (PCP).

2. Choose the correct letter, A), B), C) or D).

Listening Script

The flight interphone system is an independent communications networks. Its primary purpose is to provide private communication between cockpit crewmembers without instruction from the service interphone system. The ground crew may also use flight interphone through a jack at the external power receptacle.

The pilots can transmit directly over flight interphone by using the control wheel PTT switch. Alternately, any crewmember with an Audio Selector Panel can transmit/receive over flight interphone by using their respective ASP and normal push-to-Talk switches. Any standard microphone may be used with the flight interphone system.

Question (1): What is the flight interphone system?

Question (2): What is the primary purpose of the flight interphone system?

Question (3): How can the ground crew use the flight interphone?

Question (4): How can the crewmembers transmit over flight interphone by using the control wheel PTT switch?

Listening 4　Cockpit Voice Recorder

1. Fill in the blanks.

Listening Script

A Cockpit Voice Recorder (CVR), sometimes referred to as a 'black box', is a flight recorder used to record the audio environment in the flight deck of an aircraft for the purpose of investigation of accidents and incidents. This is typically achieved by recording the signals of the microphones and earphones of the pilots' headsets and of an area microphone in the roof of the cockpit. A flight data recorder (FDR) (also ADR, for accident data recorder) is a kind of flight recorder. It is a device used to record specific aircraft performance parameters.

The cockpit voice recorder system simultaneously records four audio channels on a 30-minute continuous loop tape. Audio older than 30 minutes is erased. Three of the channels record all receptions and transmission from the audio control panels. The fourth channel records audio from the area microphone installed on the overhead panel.

The cockpit recorder system operates anytime when AC power is available. The audio control panels are used to manage the radio, satcom, and interphone communication system. Navigation receiver audio can also be monitored. The captain, First Officer, and first observer audio control panels are installed on the aft aisle stand. The second observer audio control panel is installed on the second observer's panel. The radio tuning panels are used to tune the VHF and HF radios associated with the respective VHF and HF radios.

Listening 5   Tenerife Disaster

1. Complete the table.

Listening Script

The Tenerife disaster, the deadliest crash in aviation history, occurred on March 27th, 1977, when two jumbo jets (both Boeing 747s) collided in the fog on the ground at Los Rodeos airport, Tenerife, in the Canary Islands. At least 560 people died as a ball of fire erupted into the sky, and the explosion heard all across the island. All 249 passengers aboard the KLM plane died in the crash. About 60 injured survivors escaped the Pan Am Boeing 747, which carried a 16 crew members and 378 passengers. Neither airline was even scheduled to be at that airport, but both had been diverted from the larger Las Palmas airport due to a terrorist bomb blast there. The immediate cause was that the KLM jet was in the process of taking off and hit the Pan Am plane as it taxied across the runway. The KLM pilot was ultimately held responsible, and the Pan Am pilot cleared. The Dutch report tried to emphasize that the American plane had taxied slightly beyond the third exit, and reported that the KLM captain assumed that he was given clearance to take off due to the ambiguous terminology used at the time.

Lesson 17   Flight Control System

Listening 1   Flight Controls

1. Fill in the blanks.

Listening Script

Aircraft flight control systems consist of primary and secondary systems. The primary flight controls are elevators, ailerons, and rudder. The control column, control wheel and rudder pedals control these flight control surfaces.

The primary flight controls are powered by redundant hydraulic systems.

Secondary flight controls include a moveable horizontal stabilizer, spoilers, and leading and trailing edge flaps. Spoilers operate differentially to assist ailerons for roll control and symmetrically as speedbrakes.

2. Answer the questions.

Listening Script

Pitch control is provided by:

- two elevators   • a movable horizontal stabilizer

Roll control is provided by:

- two ailerons   • spoilers   • two flaperons

Yaw control is provided by a single rudder.

Flaps and slats provide high lift for takeoff, approach, and landing. Symmetric spoilers are used as speedbrakes.

Listening 2   Primary Flight Controls

1. Choose the correct letter, A), B), C) or D).

Listening Script

Now let me say something about the flight control systems. The flight control systems in most general aviation airplanes consist of cockpit controls, cables, pulleys, and linkages connected to the movable control surfaces outside the airplane. There are three primary and two secondary flight control systems. The primary flight control systems consist of the elevator, aileron, and rudder, which are essential in controlling the aircraft. The secondary control systems consist of the trim tabs and wing flaps. The trim tabs enable the pilot to trim out pressure, and the flaps enable the pilot to change the lifting characteristics of the wing and also to decrease the speed at which the wing stalls.

Question (1): What is the main topic of the talk?

Question (2): According to the passage, which of the following is not mentioned as a component for the flight control system?

Question (3): Which systems are essential in controlling the aircraft?

Question (4): What is the function of flaps when the wing stalls?

2. Complete the summary.

Listening Script

The primary flight control system uses conventional control wheel column and pedal inputs from the pilot to electronically command the flight control surfaces, the system provides conventional control feel and pitch responses to speed and trim changes. The system's electronic components provide enhanced handling qualities and reduce pilot workload. The primary control system design is highly redundant, with three operation modes. The primary control and stabilizers are powered by redundant hydraulic sources. Flaps and slats are hydraulically powered with an electrically powered backup system. Each of these three operation modes is controlled by specific devices built into the wings and tailplane of the aircraft.

3. Match the components (1) ~ (3) with the features A) ~ G) .

Listening Script

Ailerons control roll about the longitudinal axis. The ailerons are attached to the outboard trailing edge of each wing and move in the opposite direction from each other. Ailerons are connected by cables, bellcranks, pulleys and/or push-pull tubes to a control wheel or control stick. Moving the control wheel or control stick to the right causes the right aileron to defect upward and the left aileron to deflect downward. The upward deflection of the right aileron decreases the camber resulting in decreased lift on the right wing. The corresponding down ward deflection of the left aileron increases the camber resulting in increased lift on the left wing. Thus, the increase lift on the left wing and the decreased lift on the left wind and the decreased lift on the ring wing causes the airplane to roll to the right.

The elevator controls pitch about the lateral axis. Like the ailerons on small aircraft, the elevator is connected to the control column in the flight deck by a series of mechanical linkages. Aft movement of the control column deflects the trailing edge of the elevator surface up. This is usually referred to as up 'elevator' .

The rudder controls movement of the aircraft about its vertical axis. This motion is called yaw. Like the other primary control surfaces, the rudder is a movable surface hinged to a fixed surface, in this case to the vertical stabilizer, or fin. Moving the left or right rudder pedal controls the rudder. When the rudder is deflected into the airflow, a horizontal force is exerted in the opposite direction. By pushing the left pedal, the rudder moves left. This alters the airflow around the vertical stabilizer/rudder, and creates a sideward lift that moves the tail to the right and yaws the nose of the airplane to the left.

Listening 3   Secondary Flight Controls

1. Dictation.

Listening Script

(1) Secondary flight control systems may consist of wing flaps, leading edge devices, spoilers, and trim systems.

(2) Then leading edge slats will retract to the extended position when the pitch angle is below the stall attitude.

(3) The Boeing 737 is fitted with six hydraulically powered spoilers on both wings, which are divided into two ground spoilers and four flight spoilers.

(4) The maximum speed for each flap position is presented on a 'flaps limit placard' in the

cockpit.

(5) The leading edge devices consist of the leading edge flaps and slats and are both located on the leading side of the wing and are hydraulically powered by hydraulic system B.

Listening 4　Autopilot

1. Fill in the blanks.

Listening Script

To be sure, automation is widely used in commercial aviation and has been praised for making the skies much safer. But there are growing concerns that the industry is relying too much on automation, especially overly complex systems. Boeing, meanwhile, is facing questions about its decisions to withhold information from regulators about the anti-stall technology implicated in the Lion Air crash.

A preliminary report from Indonesian investigators indicates that Lion Air 610 crashed because a faulty sensor erroneously reported that the airplane was stalling. The false report triggered an automated system known as Maneuvering Characteristics Augmentation System (MCAS).

Once in flight, the Lion Air crew was unprepared for the automated response set off by the faulty angle-of-attack data. The pilots fought the automated system, trying to pull the nose back up. They did not succeed.

Ethiopian Airlines 302 was similar to that of the Lion Air 610. Both planes struggled to maintain altitude in the minutes after takeoff.

2. Write T if the statement agrees with the information, and F if the statement contradicts the information.

Listening Script

Autopilot is an automatic flight control system that keeps an aircraft in level flight or on a set course. It can be directed by the pilot, or it may be coupled to a radio navigation signal. Autopilot reduces the physical and mental demands on a pilot and increases safety. The common features available on an autopilot are altitude and heading hold.

The simplest systems use gyroscopic attitude indicators and magnetic compasses to control servos connected to the flight control system. The autopilot system also incorporates a disconnect safety feature to disengage the system automatically or manually. These autopilots work with inertial navigation systems, global positioning systems, and flight computers to control the aircraft. In fly-by-wire systems, the autopilot is an integrated component. Additionally, autopilot can be manually overridden.

Chapter 5   Human Factors

Lesson 18   Introduction to Human Factors

Listening 1   What is Human Factors in Aviation?

1. Listen to a speech and answer the questions as below.

Listening Script

At the university I attend in South Florida, the aviation program is quite small. Most of the courses are done online, rather than in-person and the flying lessons are done through a specialized flight school. Because of this, when I tell other students my major they seem shocked that our school has an aviation program. Even professors who have been teaching at my university for 20 years did not know we had an aviation program. One reason my major is not well known is because it is not a usual aviation major. Most majors are either in flight science, mechanics, or aviation business administration. My major is Human Factors in Aviation, and if you have never heard that before, or heard of it and don't know what it is then you are definitely not alone.

My major takes a look at a more recent development in the aviation industry: human factors. To put it in very simple terms, the course looks at the human side of aviation. In aviation nearly half of all aircraft accidents are caused by pilot error. As students of human factors in aviation, we look at what those pilot errors are, what causes them, and how we can identify them in ourselves and fix them before they become a problem. This in turn will make safer pilots and not getting ourselves into dangerous situations.

So what are these human factors? One of the biggest is fatigue. As college students we have all experienced fatigued, especially studying for finals but how does this affect a pilot? A pilot is required to be sharp and focused in order to be ready for any type of emergency that can occur during a flight. A fatigued pilot will be sluggish and slow to react to an unexpected situation. This delay in the recovery for an emergency can prove deadly. This was brought to the public light with the crash of Colgan Air flight 3407, whose pilots were found to be fatigued and not receiving enough rest.

One of the next biggest human factors is stress. Everyone feels stress in their life, including pilots. Pilots face stress from numerous sources; the weather, their company, the schedule, traffic, etc. all play a part in stress for pilots. These stresses can cause pilots to rush and skip safety steps or make a poor decision. One of the best ways to avoid being stressed is to simply not fly while stressed or if not possible, to always focus on safety when flying.

These two factors are only a fraction of the many human factors that affect pilots when flying.

Studying these factors can help make student pilots safer and make them better professional pilots in the future. Although it may not be a very mainstream major in the aviation industry, it is still very important to study these factors so that pilots can be safer in the future.

2. Fill in the blanks.

Listening Script

For years, pilot fatigue has been a real issue. Airline pilots, as well as cargo, corporate and charter pilots, can all face fatigue while on the job. While pilot fatigue can be common and overlooked, it poses a very troubling threat to aviation safety and should be taken seriously. Obviously, fatigue is caused by lack of sleep. But it's not always that simple. It can manifest acutely, such as after a runner completes a marathon, or over time, which we may know as burnout. Here are some specific causes of fatigue: Lack of quality sleep; Sleep disturbances; Interruption of circadian rhythm; Mental or emotional stress (such as family problems, anxiety, or check ride stress); Physical exertion, such as heavy exercise; Poor health, including dehydration or poor diet. Specifically, fatigue in pilots can be caused, or amplified by, the following:

Commuting: some pilots start their day 2 ~ 3 hours earlier than others to commute to work. Some have to drive a long distance to the airport; more often, though, a pilot's commute is because he doesn't live near his home base at all, and he must fly in from a different airport, adding hours to the beginning of his day.

Layovers at airports: sometimes pilots will have a 12-hour layover at an airport, where they are meant to rest. Instead, some choose not to sleep, or otherwise can't get to sleep. They watch TV, check email, or catch up with old friends and might get a few hours of sleep before their next flight departs.

Jet-lag: More apparent with long-haul pilots, jet-lag can be a big problem when it comes to pilot fatigue. Most operators give ample time for pilots to adjust to jet lag, but the body does go through stress when its circadian rhythm is interrupted, making it hard for pilots to sleep when they need to, and difficult for them to stay awake later when they need to.

Night flying: Cargo pilots, especially, deal with fatigue when flying lengthy routes at night due to the imbalance of the body's natural circadian rhythm. This will be especially true for those pilots that have varying schedules or alternate day and night shifts.

Monotonous tasks: Pilots that fly the same aircraft on the same routes into the same airports daily are prone to boredom fatigue.

3. Answer the questions as below.

Listening Script

Fatigue and sleep deprivation

Fatigue, tiredness and sleep deprivation can lower a pilot's mental and physical capacity quite dramatically. Fatigue can become deep-seated and chronic. If personal, psychological or emotional problems are not resolved, they prevent deep rest and good sleep over a prolonged period. Chronic fatigue won't be cured until the problems are resolved, or at least are being addressed, and the person can relax and unstressed. You should prohibit yourself from flying if the distress is distracting. Short-term fatigue can be caused by overwork, mental stress, an uncomfortable body position, a recent lack of sleep, living-it-up a little too much, lack of oxygen or lack of food. To guard against fatigue that is detrimental to flight safety, you should: fly only when your psychological and emotional lives are under control; maintain a reasonable level of fitness; eat regularly and sensibly; have adequate and effective sleep; ensure that cockpit comfort is optimized and that energy foods and drink are available on long flights; and exercise your limbs occasionally and, if practicable, land to stretch your legs at least every four hours.

Listening 2　Visual Illusions

1. Answer the following questions by deciding if each sentence is true (T) or false (F).

Listening Script

The eyes provide us with a visual image of theenvironment. Each eye acts as a natural and very sophisticated digital camera. Its basic function is to collect light rays reflected from an object, using the lens to focus these rays into an image on a screen. The brain then matches the image to previously stored data so we then 'recognize' the object. Seeing is both a visual and a mental process.

The central foveal region of the retina containing mainly cones is not effective by night, causing an area of reduced visual sensitivity in your central vision. You need to rely on your peripheral vision which is provided by the rods in the outer band of the retina. An object at night is more readily visible when you are looking to the side of it, rather than directly at it.

A runway that is larger than usual will appear to be closer than it really is. Conversely, a runway that is smaller that usual will appear to be further away than it really is. A wide runway, because of the angle at which you view it peripherally in the final stages of the approach and landing, will cause an illusion of being too low, and you may flare and hold-off too high as a result. This may lead to 'dropping-in' for a heavy landing. Conversely, a narrow runway will cause an illusion of being too high, and you may delay the flare and make contact with the runway earlier than expected. If you know that the runway is wider or narrower that what you are familiar with, then you can allow for this in your visual judgment of flare height an hold-off prior to touchdown.

2. Fill in the blanks.

Listening Script

Most runways are of known width and on level ground. On every approach, you should try to achieve the same flight path angle to the horizontal, and your eyes will become accustomed to this, allowing consistent approaches along an acceptable approach slope merely by keeping your view of the runway through the windscreen in a standard perspective.

If approaching a sloping runway, the perspective will be different. A runway that slopes upwards will look longer, and you will feel that you are high, when in fact you are on slope. The temptation will be for you to go lower and make a shallower approach. If you know that the runway does have an upslope, you can avoid this tendency.

A runway that slopes downwards will look shorter, and you will feel that you are low, when in fact you are on slope. The tendency will be for you to go higher and make a steeper approach. If you know that the runway does have a downslope, you can anticipate and avoid this tendency.

Listening 3   Alcohol and Flight

1. Write T if the statement agrees with the information, and F if the statement contradicts the information.

Listening Script

A pilot who was arrested for being intoxicated less than an hour before his scheduled flight from London to Tokyo allegedly had nearly 10 times the legal blood alcohol content in his system, the Associated Press reports.

On Thursday, Japan Airlines co-pilot admitted to failing a breathalyzer test on Oct. 28th and pleaded guilty to being intoxicated, the outlet reports. According to London's Metropolitan Police, the 42-year-old pilot was found to have 189mg of alcohol per 100mL of blood in his system.

The legal limit for a pilot is 20mg, while the legal limit for drivers in the U. K. is 80 mg. CNN and Japanese broadcaster NHK corroborated these reports.

London's Metropolitan Police did not immediately respond to PEOPLE's request for comment. According to a statement from Japan Airlines on their website, the co-pilot 'failed a breath test in violation of UK's Aviation Law' which revealed he had 'excessive alcohol' in his system. He then submitted a blood sample, which 'concluded that the co – pilot was in violation by seeking to fly with a blood alcohol reading in excess of the permitted level'.

According to NHK, the driver of a crew bus smelled alcohol and called police. The drunk co-pilot was scheduled to fly a 244-passenger flight from London Heathrow International Airport to Tokyo just 50 minutes after he was arrested. The local station reports he admitted to drinking two bottles of wine and a pitcher of beer the night before he was scheduled to fly the plane.

2. Listen to a lecture, and complete the table as below.

Listening Script

Alcohol

Even small quantities of alcohol in the blood impair a pilot's immediate performance and severely affect judgment. It takes time for the body to remove alcohol. A pilot must not fly for at least 8 hours after drinking the last of any alcohol. This time interval must be increased if greater quantities are consumed. You must allow at least an extra hour per standard glass.

People who have consumed large amounts of alcohol can become aggressive or engage in risky behavior. Excessive drinking will lead to confusion, nausea, vomiting and unconsciousness. In extreme cases, you may even stop breathing. Drinking copious amounts of alcohol can result in hangovers, memory loss and blackouts.

The dangers of alcohol include: impairment of judgment; release from inhibitions; slowed reaction time; impaired performance; compounding of the detrimental effects of hypoxia; reduced attention to detail; and increased probability of being involved in an accident or and incident.

Sadly, high residual blood alcohol levels have been documented in many pilots who have been involved in fatal aircraft accident.

Listening 4　Decision Making

1. Fill in the blanks.

Listening Script

Aeronautical decision making, ADM, is a systematic approach to the mental process used by aircraft pilots to consistently determine the best course of action in response to a given set of circumstances. Risk Management is the part of the decision making process which relies on situational awareness, problem recognition, and good judgment to reduce risks associated with each flight. Risk Elements in ADM take into consideration the four fundamental risk elements: the pilot, the aircraft, the environment, and the type of operation that comprise any given aviation situation. It is usually not a single decision that leads to an accident, but a chain of events triggered by a number of factors. The poor judgment chain, sometimes referred to as the ' error chain', is a term used to describe this concept of contributing factors in a human factors—related accident.

Steps for good decision making are: (1) Identifying personal attitudes hazardous to safe flight; (2) Learning behavior modification techniques; (3) Learning how to recognize and cope with stress; (4) Developing risk assessment skills; (5) Using all resources in a multicrew situation; (6) Evaluating the effectiveness of one's ADM skills.

References

[1] Aviation English (General English) [M]. Melboume: RMIT Training Pty Ltd, 2006.

[2] Henry Emery. Andy Roberts. Aviation English [M]. London: Macmillan Publishers Limited, 2008.

[3] Federal Aviation Administration. Pilot's Handbook of Aeronautical Knowledge [M]. Oklahoma: United States Department of Transportation, 2008.

[4] Guided Flight Discovery: Private Pilot [M]. Jeppesen Sanderson, Inc. 2006.

[5] 刘若莹, 马涛. 中国民航飞行人员英语: 听力教程 [M]. 北京: 中国民航出版社, 2002.

[6] 范海翔, 等, 飞行专业英语 (听力) [M]. 北京: 清华大学出版社, 2014.